While You're Here, Doc

While You're Here, Doc

Farmyard Adventures of a Maine Veterinarian

Bradford B. Brown, DVM

Tilbury House Publishers
Gardiner, Maine

TILBURY HOUSE, PUBLISHERS
2 Mechanic Street
Gardiner, Maine 04345
800-582-1899 • www.tilburyhouse.com

First paperback printing, April 2006
10 9 8 7 6 5 4 3 2

The names and other identifying details of the characters in this book have been changed to protect individual privacy.

Library of Congress Cataloging-in-Publication Data
Brown, Bradford B., 1929-
While you're here, Doc : farmyard adventures of a Maine veterinarian / Bradford B. Brown.
p. cm.
ISBN-13: 978-0-88448-279-6 (pbk. : alk. paper)
ISBN-10: 0-88448-279-0 (pbk. : alk. paper)
1. Brown, Bradford B., 1929- 2. Veterinarians—Maine—Biography. 3. Domestic animals—Maine—Anecdotes. 4. Veterinary medicine—Maine—Anecdotes. I. Title.
SF613.B76A3 2006
636.089092—dc22

2005032418

Front cover photo by Hope Millham
Back cover photo and Farm Calls photo section by Lou Garbus
Designed on Crummett Mountain by Edith Allard, Somerville, Maine
Covers printed by the John P. Pow Company, South Boston, Massachusetts
Printed and bound by Maple Vail, Kirkwood, New York

To all the animals that made it possible

To the memory of my brother, colleague, and father figure,
Philip R. Brown, DVM

To my mother, Hazel Crosby Brown

To my three children,
Jane Roy Brown,
Jeffrey Philip Brown,
and Laurah Jean Brown

To Francis H. Fox, DVM.
the best teacher I ever had

contents

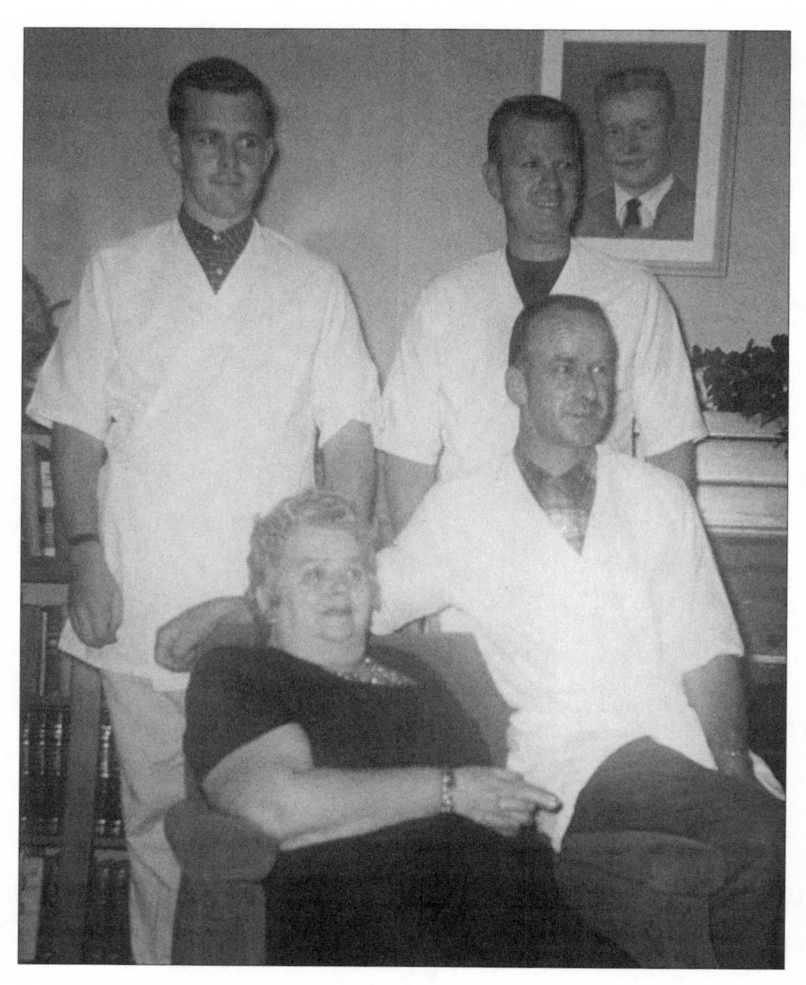

Veterinarians in the family: Dr. Neal C. Brown (standing left), Dr. Bradford B. Brown (right), and Dr. Philip R. Brown (seated), with their mother, Hazel C. Brown. Courtesy of the author.

Preface

This book is about my twenty-three years of veterinary practice, treating animals large and small, in Belfast, Maine, then a town of 5,000 people on Penobscot Bay. In the 1950s and '60s my brother Phil and I ran a small-animal hospital and made hundreds of farm calls to the dairy herds that dotted the Maine countryside.

Phil, my oldest brother, had been very generous in helping me financially during my four years at Cornell University College of Veterinary Medicine, and I joined him in 1956, right after graduating. My youngest brother, Neal, who also graduated from Cornell vet school in 1962, joined our Belfast practice briefly, but then left to pursue a Ph.D. in pharmacology at the Yale University School of Medicine. Subsequently, he became professor and founding chair of the Department of Pharmacology at the University of Massachusetts School of Medicine. Phil and I worked together until 1968, when he left the practice. I kept it up on my own for another ten years, running the small-animal hospital and making farm rounds in three counties.

My days in practice were full of physical and emotional challenges. When I joined Phil, our practice was about 30 percent small animal—the veterinary term for pets, chiefly dogs and cats. The large-animal side—horses, pigs, goats, sheep, and mainly dairy cows, in those days—involved driving to farms within a forty-mile radius of Belfast, and sometimes we'd travel as far as a hundred miles if the vet who served a distant town couldn't get there.

My large- and small-animal clients gave me very different experiences. Though farmers can be sentimental about their cows and horses and sheep, more often they treat them as a business commodity. If it's going to cost $100 to operate on a cow that's worth $400 as a milker, and the farmer can get $300 for "beefing" her—selling her for meat—he doesn't have much use for your scalpel.

Pet owners, of course, viewed their animals as beloved companions. Here I had the chance to practice medicine much the way human physicians do, with many more choices about treatment. People with a sick dog or cat would, I found, often spend their last dime to save the animal. I certainly never exploited that, but I mention it because it meant that I was able to exercise some of my skills, such as some fairly complicated surgery, and explore treatment options that the frugal farmers of that time and place could rarely afford.

This book focuses on the farm side of my practice. The farmers in coastal Maine at the time were conservative, pragmatic, and frugal. While some were well educated, many were not, and they remained suspicious of new people, new methods, and new technology. Often they had grown up without ever leaving their ancestral homestead, repeating the generational struggle against the two most unpredictable forces in the natural world: weather, and flesh and blood. Most of them tried to eke a living out of small-to-middling-size dairy herds, Holsteins for the most part, because the notoriously stony soil along the coast gave them few options to grow cash crops. Any open farmland was used to grow hay and corn silage for cattle feed.

I understood that well enough. Along with four brothers and a sister, I grew up in the 1930s and '40s on a farm in Vassalboro, Maine. A rural village of pastures and farmsteads in the Kennebec River Valley, Vassalboro was then no more than a speck on the map between Augusta and Waterville. Some people recall that it's the town mentioned in the "Bert and I" joke, when the city fella drives up to a farmer and asks how to get to Vassalboro. "Don't you move a goddamned inch," the farmer replies.

On our farm, we kept dairy cows, pigs, sheep, chickens, and horses, not to mention assorted dogs and cats. (I have no doubt that growing up in the company of so many animals was the reason why two of my brothers and I—three out of six children—went on to study veterinary medicine.) The only cash crop we raised was dry beans. The proceeds were used to pay real-estate taxes and other absolute necessities. Other cash was scarce, as my father made it clear one day when he gathered the four of us boys into the living room of our farmhouse. Our ages spanned seven to eleven, and we were somber. What had we done now? My father didn't mince words. "To get through this life, to feed yourself and make a living, either you're born rich, or you're going to have to work," he began. "I thought at this point in your lives I'd better give you the news: you weren't born rich."

That was surely true, but we had what we needed, even in the depths of the Great Depression. We lived off the land. From the ages of five to ten, I did many tasks, mostly menial, on the farm. We would still have to do all the chores, which included the milking of course, and all the chores related to the well-being of cattle and workhorses. Before the farm was electrified in 1937, my brothers and I had to haul water from the brook, a quarter-mile south of the barn. This was labor-intensive in winter, especially during a storm. We had a wooden pung—a sled made of oak with large runners, pulled by two horses—with planks placed across it to hold six fifty-gallon drums. After drilling holes in the ice, we had to fill each of those by dropping twenty-quart pails into the brook. It seemed like the six barrels would never fill, but at least the horses were getting rest before hauling those barrels straight up the bank of the brook. In a blizzard they were completely covered with snow, as were we. We hauled pail after pail, hoping there would be enough water to go around that evening so we would not have to return to the brook that night. Back in the barn, we watered and groomed the work-horses, brought fresh timothy hay from the mow and apportioned it to the cattle and forked out a generous amount to the horses. My job then was to extinguish all the lanterns in the stable and wait five minutes to make sure they were completely out. During that time of darkness I felt completely at peace with the world. The barn was warm, the cattle enjoying their hay, and all was well. At last I opened the door and ran through the blizzard to the house and a sumptuous supper prepared from 100 percent homegrown ingredi-ents. Looking back, such experiences reinforced my belief that farm-ing and nature are one.

When I reached age eleven I learned to harness the workhorses by myself and set about doing an adult's work, which I found greatly satisfying. Now I was able to join the older boys and men doing field work—plowing, planting, mowing hay, haying, and all manner of related chores.

Doing a man's work at eleven made my chest swell a bit, as I perceived I was treated more fairly by the adults in my life. But that was also a sad year for our family. My father, who experienced episodic depressions, committed suicide. That put an end to work-ing on my father's farm. My brother Phil was sixteen then, and he finished his third year at Kent's Hill preparatory school. My mother, an ardent believer in education to its highest level, made a deal with the headmaster to enable my brother to finish his fourth year by assuming two or three more student jobs. He was among the youngest valedictorians to have graduated from that school at the time. He could have been accepted to any college, but he chose the University of Maine at Orono, the parent of the state university system. The scholarships they awarded him made it the least expensive way to pursue premedical studies.

Meanwhile, my two other brothers, my sister, and I, through my mother's efforts, attended Higgins Classical Institute, a preparatory school for children of lesser means in the little town of Charleston, about twenty miles north of Bangor. As there was then no public transportation to high school in Vassalboro, students who had to travel to school from distant farms tended to drop out, as few rural families had cars then. My mother didn't want that temptation anywhere near us. So she sent us off to Higgins, one by one, and the farm basically ceased operating.

During the summers, we all started working off the farm. I took jobs with the railroad, the state highway crew, and other physical labor. None of these employments paid more than thirty-seven and a half cents an hour. But it beat working on the farm, where a week's pay came to a dollar plus room and board.

Higgins was very strong academically. For every nickel we spent on our schooling there, we got back countless dollars in the future. One brother went on to Colby, and my sister went on to Massachusetts General Hospital School of Nursing. I followed my older brother to the University of Maine at Orono, taking premedical studies in the school of agriculture. Neal, like Phil at Kent's Hill, was one of the youngest-ever valedictorians to have graduated from Higgins at that time. He also went on to the University of Maine for premedical studies. While attending college, all of us worked at least two jobs all the time. I delivered milk and worked at the ag school farm and in the pathology lab.

One of my biggest supporters in my undergraduate years was the university president, Dr. Arthur Hauck, who took a personal interest in our family. Perhaps everyone felt that way about him. He knew most of the students by name. Every semester he personally sleuthed around to see that I received all available scholarship money. Many years later, when I was practicing on my own in Belfast, Dr. Hauck appeared in my waiting room with his dog. I had the unexpected pleasure of caring for his dogs and enjoying many conversations with him during his retirement.

At Cornell, Phil graduated second in his vet school class of 1946, topped by Neal, who led his Cornell class of 1962. As for myself, I graduated near the center of my class, although I was near the top in the clinical part of our training. At my tenth vet school reunion I was walking down a long corridor when I met a former professor who had taught all three of us. He asked how I was and turned to leave. Then he stopped abruptly.

"Dr. Brown, may I ask you a personal question?"

"Certainly," I said.

"Did you and Neal and Philip all have the same parents?"

I was a little taken aback, but I knew why he was asking. "Believe it or not, we did," I replied. "But remember, Doctor: the best part of a sandwich is in the middle."

He burst out laughing. "Score one for you, Brown." He continued down the hall and I could hear him chuckling into the next corridor.

After coming up the hard way, I could hardly blame the Maine farmers who finagled to save a nickel on any service provided to them, including veterinary treatment. Many of them were scraping the poverty line and sometimes dipping below it. Their frugal modus operandi is the reason I've chosen the title *While You're Here, Doc*.

Let's say Farmer Smith, who lived about twenty-five miles from my hospital in Belfast, needed me to tend to a sick cow. Because I charged by the mile, one-way, for my farm calls, Smith knew he could divide the cost of the mileage by as many farmers as he could round up in his area. So he called all the surrounding farms to see if anyone else needed my services that day. Only then would he call and ask me to drive out to his place. Suffice to say I'd be the last person to know about the full agenda until I drove into Farmer Smith's yard, usually about twenty-four hours behind in the cases demanding my attention. Then, after the perfunctory chat about the day's weather or the low price of milk, Smith would shuffle his feet a bit and say, "While you're here, I'd like to have ya check over Mary's foot, and Jo's udder seems to be goin' sour in two quarters. I got six cows I need checked for pregnancy, and maybe you can dehorn those three calves I got tied up in the shed."

After taking care of all these extra matters, I'd be washing up when Smith would start scratching his head. "Hey, Doc, I almost forgot—Mildred says the dogs need their rabies shots. I'll try to round 'em up for ya."

Less often, Smith would end our visit with the announcement, "While you're in the neighborhood, Doc, Eli Dow needs ya, Hershel Peavey wants you to stop by for a sec, and Orville Colby says one of his cows is off her feed." To top off this list, my office would have invariably called to leave a message with him about an emergency forty miles in the opposite direction.

After Phil left the practice, I continued alone in the business we'd built together, working eighteen hours a day on average. I would tend to the dogs and cats for eight or nine hours, then, in late afternoon and evening, I'd depart for farm calls, returning home at two or three in the morning. This went on six days a week most of the time, and it wasn't uncommon for me to work the seventh day as well. The workload and the pace of this life became unsustainable. I left the raising of our three children largely to my ex-wife, who also suffered as a result of my continual absences. The exhaustion and stress of constant physical work also took its toll on my body. In 1977 my doctors advised me to slow down if I wanted to see age fifty. But I knew myself too well; slowing down wasn't an option when the phone just kept on ringing. Instead I sold the practice to two well-qualified vets.

Despite the toll it took on my health and family life, my work gave me many moments of joy and satisfaction. I enjoyed the rural people of Maine, the animals, and the many adventures on which they both took me. I hope you do, too.

Bradford B. Brown, DVM
Vassalboro, Maine

Acknowledgments

It was largely because of the generous support and encouragement of my eldest brother, Philip R. Brown, whose practice I joined after graduating from Cornell University College of Veterinary Medicine in 1956, that I chose my career. When he left the practice in 1968, he also left me with a strong work ethic and a thriving business. He managed to pack in seventy-five years of living before his death from cancer at age forty-nine. His love for his family, his profession, and his fellow man inspired me to write this book.

I also owe a longstanding debt of gratitude to Francis H. Fox, DVM, who was far and away the best teacher I had in twenty years of formal education. For generations of large-animal veterinarians, he provided unexcelled knowledge and inspiration at Cornell's vet school, and he remains a legend there. He was my guide, confidante, and mentor during my four years in the veterinary medicine program. Without his confidence in me I may never have continued on to the adventures that form the basis for this book.

Finally, I'd be remiss not to thank my eldest daughter, Jane Roy Brown, whose encouragement, transcription, and editing have helped extract the book from my mind and get it onto the page. Without her help I doubt this book would exist.

1

Stormy

It was about twenty below zero and snowing heavily at 4:00 A.M. as my car struggled to get to Herford Hutchins's farm. By the time I entered the driveway, at least 10 inches of snow covered the ground and the radio forecaster was predicting another six inches by mid-morning.

Herford had called a few hours earlier to say that his new stallion had caught his ear on a spike in the stall sometime during the night, and the ear was hanging by a flap. Finally I reached his farm and struggled up the unplowed driveway, parking as close to the barn as I could. Bundled in a green-plaid wool jacket, Herford came out of the barn to greet me. "Doc, if this keeps up we'll get that storm they're talking about," he said.

"Looks like it," I said.

Herford was in his mid-thirties, an exceptionally quiet person. On the rare occasions when he did speak, it was very softly, and he barely parted his lips in order to keep his snuff tucked under his lower lip. Though only five feet seven inches tall, he was as rugged as a wrestler and hadn't an ounce of fat on him. He kept a few dairy cows and worked in the woods every day. Some people said he was half-horse, such was Herford's love for the animals. The problem was, he tended to collect "outlaw horses"—those with high tempers and bad habits, such as biting and kicking people. Especially me.

"He woke me up," Herford said, as I popped my trunk and pulled together the things I'd need to sew up the horse's ear. "I thought the barn was coming down, the way he was thrashing and pounding. Stormy, he's got a terrible mean streak in him. He's man-shy and hard to handle, too."

Many a horse owner had told me similar things over the years, and often the animals turned out to be manageable if handled with care. But the same words coming from Herford meant our work was going to be hazardous at best. "Stormy, eh?" I said. "Well, he sure picked the right night to get into trouble."

We trudged through the snow to the barn. Stormy stood facing the wall of his stall, which had a door made of wooden slats. He wore a rope halter that was tethered to a hook on the left wall. Smelling and hearing us, the horse snorted and stamped his feet.

I stood at the closed door of his stall and shined a flashlight on the right side of his head. The beam revealed a bloody mess. The skin covering the right ear was hanging by a thread, and the ear cartilage was ripped along its entire length—a painful injury. To compound his problem, the poor beast had ripped his right upper eyelid on the spike, and it had dropped below the bloody surface of the right eye, which meant he couldn't see out of that eye at the moment. If we were going to treat him, I needed to get at his head.

"What do you think, Herford? Can you get him out of the stall?"

Herford considered this for a moment. "Well, I can't use a bridle on him 'cause of that torn ear, and he's gonna be extra hard to handle on account of he can't see out of that right eye. Best not to go in the stall with him. He's a crowder and a biter, and he strikes fierce with his front feet. I only just got so I could do a little somethin' with him. It's a damned shame he's gone and done this."

This wasn't the answer I was hoping for. "Well, Herford, we have to give him a shot to knock him out, so we've got to get him out onto the floor somehow."

Herford nodded and reached through the slats. After several tries, he managed to free the horse's halter rope from the hook and swung the stall door open. Stormy backed out like he was shot out of a cannon. Herford and I retreated, pronto. Even so, we were all but annihilated before I caught Stormy's dangling halter rope. Then the panicked horse snaked me this way and that until Herford secured another rope onto the halter ring. The the three of us did a scuffling dance out onto the barn floor. Our plan was to use the two ropes to control him by hand, allowing us more freedom of movement, rather than to tether him to a post or a hook.

After leaving the office I had taken what medicine I had anticipated I would be needing and placed it on the heater vents of my car to prevent it from freezing. Before entering the barn I had stuffed every bit of medication into my pockets, including syringes filled with tranquilizer and local anesthesia, hoping to avoid having to use a general anesthetic. With an outside temperature of twenty-five degrees below zero, the air in the unheated barn would be well below zero, making it a poor operating room. Even hot water freezes in a matter of minutes, as do most liquid medications, to say nothing of my bare hands. Herford had brought a pail of hot water into the barn as he waited for my arrival, but it had started to crust over with ice. I picked up the syringe filled with tranquilizer and started to inject the fluid into Stormy, but the fluid froze the second it entered the metal needle and never made it into the horse. The pinch of the needle didn't do much for Stormy's mood. I sent Herford into the house for more hot water as I struggled to control the hysterical stallion. Herford returned in about five minutes, saying it was a mite colder out now and still snowing.

I dipped the tranquilizer syringe into the hot water until I could squeeze a test drop out of the needle. After another big struggle, we finally managed to get some tranquilizer into Stormy's neck muscle. Hoping it would calm him down long enough to inject a general anesthetic, we continued our tussle. Our luck ran out completely though when he met the barn door with his buttocks. His twelve hundred pounds split the door, throwing Herford and I violently to the floor. Stormy found himself free in the snowy barnyard and took off at a gallop into the swirling blizzard. We struggled to our feet and ran after him.

By now, a foot of snow had accumulated. It was nearly impossible to hold our heads up and face directly into the lashing wind. Even if it had been daylight, we couldn't have seen more than ten feet ahead. The subzero wind chill factor didn't help our search. Blowing snow covered our tracks as soon as we took the next step, and had also erased our fugitive's trail. We both had flashlights, and when we reached the road—the horse's logical escape route because the snow wasn't as deep—we decided that Herford would go down the road in one direction and I'd go the opposite way, hoping that maybe one of us would get lucky. Before leaving, I grabbed a bottle of anesthesia and a syringe, just in case the tranquilizer had caught up with Stormy.

Arriving at the nearest house, I noticed a porch light glowing dimly through the snow. As I knocked on the door, I saw Spurl Parker in the kitchen, pulling on his boots to head out to the barn for morning chores. I knew Spurl from many calls to his farm.

"For God's sake, Doc, what are you doing here—come on in and get warm," he said. I entered after brushing the snow from my clothes. "It's a mite balmy out there, isn't it?" the farmer said.

"Sure is, Spurl." I quickly explained the situation, and he offered to join the search.

"Y'know, Doc, when I flipped on that outside porch light about ten minutes ago, I thought I saw a moose headin' towards the village," Spurl said. "Now I know it wasn't a moose. Let me see if my old Jeep will start. Maybe we can plow our way up to the village."

Spurl went into his garage through the side door, and after a few tries was able to get his vehicle started while I shoveled snow away from the garage door so he could back out. Even with four-wheel drive we had a hard time reaching the village, sliding into snowdrifts as the intense cold and wind turned the falling snow into instant concrete. As we passed the general store, which was illuminated by a light from a neighboring yard, I noticed that the big front window had been shattered.

"Pull her over, Spurl," I cried. "The store window's been smashed."

He slowly turned into the parking lot. "Probably a branch got flyin' around in the wind," he speculated. "I'll keep 'er running and you go ahead and take a look."

Wading through the knee-deep snow, I reached the broken window and cast my flashlight beam into the store. Lo and behold, there was my patient, eating peanuts from the barrel down at the far end of the aisle. From the glazed look in his good eye, it was apparent the tranquilizer was taking effect; but meanwhile, young Stormy had had some fun. Canned goods were scattered all over the floor, and food racks had tipped over, leaving debris and broken bottles everywhere. A hand grenade couldn't have done more damage. Luckily, his motor was now running down and he seemed not to care that his pursuer had caught up with him. After a looped glance in my direction, he snorted once and plunged his muzzle back into the peanut barrel.

I flashed my light on and off into Spurl's Jeep window. Leaving the engine running, he slogged over to where I stood in front of the window. I shined my light down the aisle. "For God's sake, Doc—who'd a believed it!" he exclaimed. "What do you s'pose we ought to do?"

"Got a rope in that Jeep?" I asked, shouting against the wind.

He nodded and retreated toward the vehicle, returning a few moments later. Working quickly, we entered through the smashed window, barricaded the opening with shelves, and fashioned a crude lasso out of Spurl's rope. After several futile tosses, Spurl finally got it over Stormy's neck and secured the other end to the chimney, which gave me time to draw a syringe full of anesthetic. Thanks to the tranquilizer, the horse allowed me to get the injection into his jugular vein. Seconds later, he keeled over onto the littered floor. The quarters suddenly became very cramped, as Stormy sprawled all over the available floor space.

"God, Doc," Spurl muttered. "Never s'posed I'd see a sight like that. He went down like he was poleaxed."

"Spurl, we've gotta hurry," I said. "See if you can get down to Herford's and bring the two black grips out of my car, and grab a horse blanket out of the barn. Keep an eye out for Herford—he was walking in the other direction, but he might have gone back to the house."

Spurl waded back to the Jeep and spun away into the blizzard. Minutes later, the storekeeper, Harley Hartland, came into the store. He lived only a couple of hundred yards away and had walked over to open for business, as usual.

"Holy cow—I mean holy horse! What in hell happened? Did you ride that fella in here, Doc?" he exclaimed. Then his eyes took in the full extent of the damage. His face sank, and a long sigh escaped. "Well," he said, "I guess it could be worse—I coulda been robbed."

I hastily told Harley the story, stressing that time was of the essence.

He nodded grimly. "I'll see if I can get something to tack up over that broken window. It's gettin' colder than the hinges of a

Siberian door in here."

I monitored the horse's vital signs and grew increasingly anxious as more than thirty minutes passed. The temperature rose to about zero, thanks to Harley nailing a couple of old army blankets over the window and firing up the wood stove. I began to wonder if Spurl had made it to Herford's, which in good weather would have been less than a five-minute drive. Just as the stallion began to stir—a sign that the anesthesia was wearing off—Spurl and Herford showed up with my grips.

Herford spoke first. "I hadn't intended to do it tonight, but while you're here and got him under, why don't you cut him? Might make him more sensible."

He was asking me to castrate the stallion, and I agreed that it might make the horse less feisty. "Sure," I replied, "but we have to move fast. I only have a little anesthesia left, and I'm not sure it'll be enough to keep him under."

After asking Harley to run a clean bucket of hot water for me, I injected more anesthesia, and Stormy slipped back into deep unconsciousness.

Harley returned shortly with a metal pail full of snow. "Pipes are froze, but I fetched this and we can heat 'er on the woodstove," he said.

"Good enough," I said.

A few minutes later I poured some disinfectant into the tepid water, dropped my scalpel, needle, and suture material into it, and daubed the horse's scrotum with isopropyl alcohol—the best surgical prep I could do under the circumstances. Meanwhile, the men cleared the pickle and sauerkraut barrels and a dozen other items out of the way to make room for surgery.

I set to work on Stormy. After an hour or so, I had reconstructed his torn ear and eyelid and removed his testicles, with Spurl and Herford assisting and Harley peering over their shoulders.

After I had stitched up Stormy, I noticed Spurl staring down at the testicles. "What makes one so much bigger than the other?" he inquired. Sure enough, one was almost twice as large as its companion. I had scarcely noticed in my concentration.

"It's not unusual for two testicles to be slightly different in size,"

I said, "although I admit there's an exceptional difference in this case. Now that you mention it, it's got me curious, Spurl."

I placed the larger testicle on top of a barrel of sauerkraut and sliced it into fifty or more pieces, searching for such abnormalities as tumors or cysts that might account for the gross difference in size.

"Maybe he drew on that one less, and it just stayed fuller," Herford suggested.

"Heh," I chuckled. "That's not quite how it works, Herford. But I don't see anything abnormal that would explain it. Probably just a hereditary oddity."

Harley, who had been restacking the canned goods, asked what we were up to, and Herford told him. Harley gazed at the neatly sliced testicle and asked, "Can you eat that?"

"What!?" Herford and Spurl cried in unison. "For Chrissake, Harley, what a question."

"Actually, it's a good question," I said. "In the West, Rocky Mountain 'oysters' are a delicacy, and they're really sheep testicles."

"Yuck," Spurl said with a grimace.

I straightened up from the barrel and gazed down at the sleeping horse. Pretty soon he would be coming to. "Well, boys, how are we going to get this fella home?"

They tossed out a few ideas, and in the end, we had to accept the reality that even if we could get our hands on a vehicle large enough to haul him, we'd bog down on the snow-covered road. At last Herford said, "By God, Doc, if he walked up here in the damned blizzard he can walk back in the damned blizzard."

Meanwhile, Harley had phoned the town plow driver, whose wife reported that he'd broken down shortly after midnight and had no idea when he'd get the road cleared. She thought he'd have to hire larger equipment, because this storm was too testy for his plow. This was not good news for our heroes.

About that time Penson Pearson came wandering into the store. Pense was the town drunk in about three different towns. The last person to see him sober was the jailer in the next town over, which had the nicest jail—Pense's home away from home. Today he looked like he'd been on a bender for a month, and I know we were all thinking the same thought: How in hell did he get here? But

Pense had a knack for being where the action was.

"Gimme a six-pack of beer and one of them whoopie pies, Harley," he brayed.

Harley handed him the goods, shaking his head. "Now, listen, Pense, you can't—" Harley started to say.

Pense popped open a can. After hogging down the whoopie pie in one gulp, he tipped back his head and poured the beer down after it.

"—drink that beer in my store," Harley finished. Pense belched and crumpled the empty can in his fist. "Goddamn it, Pense, it's a state law."

"Hell, Harley, ain't no liquor inspector gonna be in here on a day like this."

Harley sighed. The horse began to stir, and our attention shifted back to him. Only then did Pense seem to notice the animal stretched out in front of him. He jumped back, stumbling. "Whoa—what the hell's goin' on? There's an elephant in here!"

Harley smiled slyly. "You're seein' things again. Last week it was snakes, now it's elephants." He shook his head.

Pense paled and staggered out the door, beer in tow. I asked Harley to give the plow driver's wife a call after we left and tell her husband to keep an eye out for Pense. He nodded. Spurl went out to start his winch on the Jeep, returning a few minutes later with a rope, which we secured around Stormy's chest like a harness. With a great deal of effort, we slid an old army blanket under the horse's torso. Then Spurl slowly reeled the prostrate animal down the aisle of the store and into the dooryard while we three pushed him along, somehow managing not to stave up the store any more than it already was.

Once out in the cold, Stormy staggered to his feet. Still tied to the Jeep, with Herford and I holding either side of his halter, he floundered through the deep snow. But soon he found steady footing. We unhooked him from the Jeep and set out on our journey, Spurl carving a path with the Jeep, Herford and I walking on either side of Stormy. Herford clutched a long rope tied to the halter, while I gripped the rope halter itself. The blizzard still raged, but it was daylight now, which helped us navigate through the drifts towards Herford's. The wind froze the hair of our nostrils, and saliva turned to rime on Stormy's muzzle. We kept our talk to a minimum.

Only about a quarter of the way there, the Jeep sank into a huge snowdrift, and Spurl joined us afoot. The horse seemed too stunned by the weather to give us any trouble, and the trip went pretty smoothly, all things considered. But as we got to the barn door, Stormy turned back into Mr. Hyde. He reared up and lashed out at the door, pounding his front hooves against the wood. His eyes rolled wildly. I released the halter and jumped back. Herford let out a lot of rope. His son Perly, who had been watching for us from the kitchen, came out to lend a hand.

"He's got a fear of doors, Doc," Herford said, shaking his head. "Always spooks goin' through this one. You and Spurl get a rope behind his rump. Perly and I will bring him forward toward the door. I'll stay on his head."

We took our positions. We had him halfway through the portal when he reared up and tumbled over backward onto the windshield of my car. Luckily it was padded with about ten inches of snow, which prevented further damage to Stormy—but not to the car. The windshield was shattered and the hood was squashed into the motor, but I couldn't take time to inspect the damage. Finally someone had the brilliant idea of blindfolding the horse, and we managed to coax him through the door and into his stall, where he settled down. Even in the intense cold, I started to relax for the first time since this adventure began.

Perly ran to get a space heater to warm Stormy's stall. Herford went to the shed and emerged with a snowmobile suit for me. Since my car was out of commission, he told me he would drive me back home on his snowmobile. In a matter of minutes, we were flying over the snow, making the fifteen-mile ride to my office in about an hour.

A week or so later I stopped into Harley's store for lunch as I frequently did when making farm calls in the area. "Hi, Doc," he said cheerfully. "I saw Herford yesterday. Says his insurance is gonna pay to fix the damage to the store and pay you for repairing your car. Stormy's healing up good too." He peered out the newly replaced window. "I see you're still drivin' it all stove up."

I nodded. "Yup," I said, perusing the contents of his deli case. "Just haven't gotten around to fixing it yet."

"While you're here, Doc, try that sauerkraut." He handed me a plastic forkful of the stuff.

"Damn, Harley, this is good—what's your secret?"

"Well," he said, "I guess you could call it Eastern Mountain oysters." I coughed involuntarily. "Boy, I'm telling you, this batch has been sellin' real good now that the brine's soaked through it. I'll give you some to take home with you, Doc."

"Uh, thanks, Harley." I could hardly refuse.

The next summer Herford brought me a picture of Stormy with his daughter riding him in the Fourth of July parade. Down in the right hand corner he had written "To Doc Brown, the coolest vet I've ever seen."

"Coolest" was underlined.

2 The Hill Farmers

It was 1:30 on a February morning when the phone rang. It had been snowing since early evening and I'd had trouble getting home from my farm calls. I had prayed I wouldn't be called out again until the wild storm subsided. No such luck.

The caller was a neighbor of the Hodge family, who lived on the side of Frye Mountain in Montville, about a dozen miles away. The neighbor told me they had a horse down in awful pain and wanted me to come as soon as possible. One of the boys would meet me at the foot of the mountain with a horse and sleigh to take me the rest of the way. I said that the boy should wait at least an hour and a half before meeting me, as the drive would take at least that long, if I could get there at all.

I took off immediately, making good time for the first twelve miles on Route 3, a major highway. When I had to turn off to a secondary road into hilly country, however, I encountered three-foot-high snowdrifts all over the road, and blowing snow cut the visibility to almost zero. I made it up the first steep hill, though it cost me half the tread on my snow tires. Before heading downhill, I rolled down my window and stuck my head out to see better, then edged down into the valley, barely making it over a one-lane wooden bridge. Because of a raging wind and a minus-twenty-degree wind chill factor, the snowdrifts blocking the road had packed down as hard as cement, and I could roll over them without getting mired. The actual location of the road was anyone's guess, so I just snaked around, up, and over the drifts. When I reached a familiar patch of thick woods, I estimated that I still had five miles to go, but it would be the roughest stretch, traveling up and down a series of hills.

As I started up the first big hill, I noticed a giant object in the path ahead. Not wanting to lose momentum on the upgrade, I pressed the pedal to the floor. The object loomed closer, and I realized it was a snowplow truck. I made a split-second decision to go around it into the field and was succeeding pretty well. But as I passed the truck, I noticed it was stopped, and the fellows in the cab were waving frantically. I got out and went over to them. Through the blowing snow I recognized Francis Lally and Alton Doucette.

"Having trouble, boys?" I asked.

"We sure are, Doc, and we're some tickled to see you. We're about half-froze. This rig broke down about an hour ago."

"What happened?"

"Well, we were just barely making it through Muzzy Gap," Fran replied. "Spent a lot of time buckin' them drifts up through there. Only made about twelve miles in six hours, and then the engine just quit. The gas line's froze solid. What the hell're you doin' out here?"

"Got a call up at the Hodge place on Frye Mountain. They have a horse that's in bad shape."

"Doc, I don't believe you're gonna make it. It's an awful bad stretch up through there. You're the first human being we've seen here for about five hours."

"Well, I'm gonna give it what I've got."

"They're kind of an odd actin' bunch up there, Doc, I'll tell ya that. Live just like hermits," Fran said.

"Cripes," Alton chimed in, "they look like a walking ad for Smith Brothers cough drops. And as far as I can tell they never learned to talk."

"Well, they talk, but they don't waste many words, that's for sure," I said. "Why don't you fellas hop in my car and warm up a bit."

"Thanks. I think we can thaw that line on your heater vents. Then maybe you could try givin' us a boost, 'cause I wore out the battery tryin' to get her started," Fran said.

The men piled into my car with the gas line. It thawed after a few minutes, and they managed to get it back into the truck. I jockeyed my car into position for the jumper cables, and after a couple of long churnings, the truck engine fired up. They waved, and I proceeded on my way to the Hodges.

There were several times when I thought I couldn't make it another foot, but inching along, I finally arrived at my rendezvous. Through the blowing snow I could see a heavily bundled draft horse hitched to a sleigh containing a single person. On closer inspection, I wasn't so sure it was a person: wrapped in a buffalo robe, he looked more like a giant gorilla. His black-bearded face and long black hair reinforced that impression. I got out of the car and approached him.

"Sorry to make you wait, son. How long have you been sitting here?"

He never uttered a sound, and it was so hard to discern his facial expression through all the hair and fur that I had to study his eyes for an answer. They rapidly expressed the fact that he'd been there too long. I grabbed the grips that I thought I'd need and clambered up onto the seat beside my driver. Still silent, he turned the horse toward the mountain road, reached under his seat, and came up with another buffalo robe, which he placed on my lap. I covered as much of my body as possible.

As we made our way up the mountain I asked the young man how the patient was doing and what had happened, trying to gain some history. But he stared straight ahead as though I hadn't spoken. It seemed like an eternity before we finally reached the Hodges' rustic abode. Upon arriving I jumped down from the sleigh, removed my grips, and the man escorted me wordlessly into the kitchen. There Dora Hodge, far more congenial than her son, introduced herself and invited me to warm myself in front of a cavernous fireplace. I looked around, feeling like I'd traveled back in time.

The room was right out of the early nineteenth century. The huge fieldstone fireplace contained cast-iron Dutch ovens and caldron hooks. Candles illuminated the room, and the slate sink had no running water. A large ham, no doubt home-cured, lay on the sideboard next to the sink. Dora noticed me glancing around and told me her sons carried all the water in from the well. She added that her husband and sons had forged all the metal utensils and hardware right here on the farm—cookware, latches, sleigh runners, wagon wheels, saws, horseshoes, virtually everything. She made the candles and soap from tallow. And except for wheat and certain other items the Maine climate couldn't support, the family raised all its own food.

I was just getting warmed up when another of the Hodge sons came in, his beard white with frost. Without a word he walked to the sideboard, ripped off a piece of the ham with his hands, and devoured it noisily over the sink. I was grateful he had turned his back to us.

Eager to get to my patient, I announced I'd better get out to the barn. At that point the young man turned around.

"Pa shot that horse half an hour ago," he said through a mouthful of meat. Turning to his mother, he added, "Don't worry, Ma, it was over in an instant." Pieces of ham clung to his beard.

Just then the kitchen door flew open and a giant figure entered in a cloud of blowing snow. It was unquestionably the man of the house, Eulie Hodge. Without acknowledging me he strode over to the ham, tore off a piece, then turned around. "Vet'nary Brown. Didn't reckon you'd make it up here. I'm some surprised you did."

I nodded. The family's taciturnity was rubbing off on me.

"About that horse. He was sufferin' somethin' wicked. I put a bullet through his head to end his misery. But since you're here, I want ya to look at our best milkin' goat. She went off her feed a couple days ago and started slobberin' outta one side of her mouth. She tries to eat a little bit but don't swallow too good. She's fallin' away to nothin'."

"Sure," I said, "where is she?"

Eulie and his second son led me through the blizzard to the goat shed. By now the snow was thigh-high and the visibility nil. This was the kind of storm where people get lost between house and barn and aren't found until the spring thaw.

Keeping my neck turtled and my eyes trained on Eulie's tracks, I managed to make it to the goat shed. Inside, I asked the son to hold my flashlight while I held the goat's mouth open and peered inside. After a careful examination, I thought I saw a foreign object in the back of her throat. To investigate further, I'd need to inject her with anesthesia. Reaching into my grip, I found that all my liquid drugs were frozen solid. I asked the son to fetch the teakettle from the kitchen and bring it back full of hot water. When he returned, I immersed the bottles I would need. They thawed quickly, and I instructed the young man how to hold the goat while I injected the anesthesia into her jugular vein. As the drug took effect, the goat crumpled in the boy's arms. "Pa," he said, "I think she's dead."

I assured him that she was okay. After laying her down on the hay-strewn floor, I checked her vital signs and proceeded to search her throat, with Eulie and his son holding her jaws open. My flashlight revealed a twig lodged in the epiglottis—a painful and eventually fatal obstruction. Barely able to get two fingers down her narrow throat, I grasped it, only to find that the twig was a piece of raspberry bush, which was studded with tiny thorns. I couldn't just pull it out, as it would tear the trachea. I explained to Eulie that the only way to save the goat would be to make an incision in her trachea, aiming for a point below the raspberry shoot, and try to extract it that way. He nodded.

"Do what you have to," he said. "She's our best milker."

The problem now was that I needed some instruments from the trunk of my car, which was at the foot of the mountain. It was just another reminder that you can never prepare for these while-you're-here situations. With no other choice, I asked Eulie for a pair of long-handled needle-nose pliers and a tin of adhesive tape, stressing that it was the container, not the tape, I needed. He dispatched his son to get them, and the young man returned ten minutes later.

Miraculously, given that virtually everything on the farm was handmade, the Hodges had a roll of adhesive tape in its original metal container, the kind with a hole in the center. The container consisted of a hollow metal spool that held the tape and a tight-fitting lid to keep the tape's adhesive from drying out. The spool's core was about an inch in diameter and an inch and a half thick—a pretty good substitute for a tracheotomy tube. I removed the tape from the spool and plunged spool and pliers into the hot water.

The trickiest part of the surgery was figuring out where to cut. After checking her throat again I made an incision and inserted the hollow spool. The tube kept the incision open while I probed the animal's trachea with the pliers. Peering through the center opening with the flashlight, I was relieved to spot the end of the raspberry shoot just above the incision. I inserted the needle-nose pliers through the spool and carefully grasped the end of the twig, pulling it gently. Slowly, a five-inch shoot emerged. It must have been coiled inside her trachea, cutting like a nest of tiny knives and obstructing passage of solid food. I mixed up an antibiotic solution and swabbed out the inside of the trachea with a wad of cotton. It didn't take long to suture the incision.

"Doc," Eulie said, "where she ain't woke up yet, why don't we trim her feet a mite."

"Well, I don't have my hoof knife, so I'll do my best with this," I said, taking out my pocketknife. I finished just as the goat started coming out of anesthesia. She soon gained her feet and within a few minutes started to nibble a few spears of hay. We saw immediately that she was swallowing a lot better, though her throat would remain sore for days.

"Golly, Pa," the Hodge boy exclaimed, "I never supposed I'd see anything like that." Then he turned to me. "If I ever have to have anybody cut on me, it's gonna be you."

"Well, I hope you never have to be cut, but if you do you'll have to use a people doctor because I don't have a license to cut people."

"Pa calls me a jackass most of the time, so maybe that makes it legal."

We all chuckled as I picked up my gear. Eulie suggested we go into the kitchen to get warm and have a cup of coffee. The storm was abating as we made our way to the kitchen.

"I ain't got a heck of a lot of money right now, dead of winter and all," Eulie told me as we drank the coffee. "But it won't be long afore I get a load of logs outta Searsmont. Soon's I do I'll write you a money order for what I owe. Just leave your bill here."

I scribbled out a slip and left it with Dora just as the taciturn son appeared to ferry me back down the mountain. The trip was mighty cold, and the drive to Belfast, though less formidable, was still a battle. As I drove into my office yard a look at my watch told me I had been gone exactly eight hours. Since my bill had come to $20, I computed my hourly wage at $2.50 an hour—a dollar less than the lowest-paid employee at my animal hospital at the time. But that was the way things sometimes worked out in my profession, and it seemed to even out over time.

In any event, Eulie Hodge made good on his word: I received payment exactly when he promised.

3

Perky Perkins Government Nemesis

Beula Perkins, called Perky by those who knew her, lived up to her nickname. She was a lean, petite, feisty woman in her late sixties who wanted only to be left alone to work her well-kept dairy operation, which she had maintained all her life. There seemed to be no end to her physical stamina: she cared for thirty dairy cows and made a home for her invalid husband without any outside help, except for a few high-school kids she hired to cut and bale hay in the summer.

Perky was all business, with no tolerance for people less ambitious than herself, and she wasn't shy about voicing that opinion. Like a lot of rural people, she was suspicious of everyone who came to her place—especially anyone who worked for the state or federal government. With a multitude of government agricultural programs involved in every aspect of farming, officials of one kind or another were periodically required to stop by, and Perky used these visits much as athletes use a workout: to exercise her healthy skepticism.

Having witnessed her in action, I am grateful never to have stepped onto her premises as a tax-paid employee. I don't recall a visit to her farm when she didn't rant about the "lazy fools" who worked for the government wasting her tax dollars. "There isn't one good day's work in the lot of 'em," she was fond of saying. "Look at 'em, speeding all over creation, wasting gas in new cars and big trucks that I paid for."

One day when I arrived she was sputtering about the new state milk inspector, who had just departed. "I taught that little feller a lesson he'll never forget," she announced. "You can't imagine what he wanted to do, Doc—move them lights over that milk tank back three feet because they might *explode* and the glass *might* get in the milk *if* the cover happened to be open!" She literally stamped her foot in anger. "Now I said, 'Young fella, I had them lights just where you want 'em, and the last inspector just made me move 'em over the tank so I could see to wash out the milk tank better. Well, they ain't about to be moved again. You got that straight, young fella?'"

Apparently the "young fella" had had the audacity to stand his ground, citing a new state regulation, and ordered Perky to move the lights before his next inspection. "About the time he opened up the cover on my full tank of milk and started runnin' his dirty fingers around the clean edge inside, he ended up head-first in that tank. I slammed that cover down on him and turned the agitator on for a minute to let him know I meant business."

"You don't say," I said. Poor devil, I thought.

"You bet your blankin' booties. He threatened to sue me. I didn't even let him use my hose to wash off. I just told him to get off my property and thank his lucky stars I didn't drown him."

Such was a typical Perky rant.

One autumn afternoon she called more agitated than usual. "Doc, how soon can you get over?"

"Probably not for a couple of hours, Perky. What's wrong?"

"I can't be saying over the wire, but be sure you get here before dark," she commanded.

When I drove into her dooryard, Perky shot out of the barn and stormed over to my car. I could see her lips moving long before my ears picked up her voice. "Get your knives and come with me," she barked. "I got two dead cows and a lot of sick ones. And I think I know what's causin' the trouble."

We headed into the pasture. After walking for two or three minutes we stopped at a crest overlooking the back slope of a twenty-acre pasture.

"Now, Doc, tell me what you see," she said, hands on hips.

Hundreds of thin aluminum strips, ranging from six to eight inches in length and about three-quarters of an inch wide, covered the pasture. In the midst of them lay two dead cows. A brilliant sunset reflected off the aluminum, giving a surreal cast to the scene of destruction. "It looks like a giant Christmas tree was run over by a steamroller," I replied.

"Where in hell did that stuff come from, Doc? My neighbors nor nobody I know would ever put that stuff in my pasture, that's for sure. I turned the cattle out here this mornin'. When I went out to get them for milking tonight, this is what I found. And nobody can tell me that stuff didn't kill them."

"Well, Perky, that aluminum must have come from an airplane during a training mission, probably launched from Dow Air Force Base. They use aluminum strips like these to jam radar in other planes. I'll bet that's what happened. They usually fly over here on their approach to Dow and must keep a record of all flights to and from the base."

"I bet you're right," she said. "Them foolish airplanes fly over this farm day and night—mighty low sometimes. There was a pile of them goin' over last night. That's when they must've dropped that stuff, 'cause it sure wasn't here yesterday. The government's gonna pay for this foolishness."

"Let's go back to the barn and check those sick cows, and then we'll come back out here and do a postmortem on the two less fortunate ones," I said.

After surveying the herd, we found at least nine more affected cows. Each was swollen with lethal methane gas trapped in her rumen—the second of a cow's four stomach compartments—a condition called bloat. I quickly released the gas by injecting local anesthesia in their left flanks and driving an eight-inch post-mortem knife through the skin and muscles and into the rumen. Then I sutured about a foot of stomach-tube hose into each cow to release the gas, saving their lives temporarily, at least. Perky and I then trekked back to the pasture, where I opened up the two dead cows. Sure enough, in each cow, the abomasum, or fourth stomach compartment, was filled with the strips, blocking the pyloric valve. The plugged valve caused digestion to grind to a halt, leading to the buildup of lethal gas in the rumen. With conclusive evidence in hand, we went to the house and phoned the air base.

"Hello," Perky said into the receiver, "I want to speak to the boss of the flyin' place. What'd you say again? I don't care if he left for the day hours ago. Give me his home phone. It's a serious emergency. No, I don't want to talk to anybody but the boss of that place.

What's that number again?" She jotted down a number. "Got it."

With that she hung up and rang the number the base had given her. "Hello, you the boss of that flyin' place? Good. You're just the feller I'm lookin' for. I got two dead cows here already and lots of sick ones 'cause your government flyin' machines dropped aluminum all over my place last night. You'd better get yourself down here and see for yourself what you done to my cows. In the mornin'? Like hell. You'll get over here tonight while these carcasses are still warm. I want you to see 'em as quick as you can so you won't have no excuses later on."

Apparently the party on the line saw that it was futile to dissuade her and agreed to come right away. Giving him directions in Maine fashion took another few minutes—"go down past where they cut down a big elm tree last spring and continue a spell till you see a rock pile on your left" and so on.

Perky and I went to the barn to continue treating the cows until he arrived. Several more critters had started to bloat, so I inserted four more tubes. When those were done we discovered bloat beginning in six more.

About an hour passed before the headlights of three cars shone on the lonely road leading to the farm. Perky and I went to the dooryard to greet them. The first person out of the car came over and introduced himself as Colonel Hackett, a doctor in charge of medical services at Dow Air Force Base. "Something came up, and the commander couldn't come himself," he said. "He thought it might be a problem I could solve."

Five other officers got out of the other two cars and came over to join us. Colonel Hackett introduced Perky to the deputy commander, Colonel Sherman. She lost no time lighting into him: why did they have to bring three cars when six of them could have ridden down in one, and there was no need for six people anyway? Perky allowed it was just typical of the government—five people to do one man's job.

Colonel Sherman ignored her diatribe and informed her that Colonel Hackett was a trained pathologist. "You don't need no pathfinder with ya, fella," she said, jabbing her finger at him. "I'll show you the way to the trouble you people caused me."

Rounding up flashlights, we all tromped to the pasture in

silence, save for Perky's occasional mutterings. "Now, Mister," Perky said when we got to the aluminum-strewn pasture, sweeping her arm out to take in the miserable scene, "you just take a look at that. Some of your airplanes dropped that stuff, sure as I'm standing here."

While the other officers strode around examining the metal strips, putting as much distance as possible between themselves and Perky, the pathologist and I examined the dead cows. The pathologist, being unfamiliar with the eating habits of cows, was incredulous that they had eaten something as obviously indigestible as shredded aluminum. I gave him a brief overview of the eating habits of ruminants, explaining that they don't chew their food before swallowing. As long as they can get it down, they eat it. In one case I learned about in vet school, a cow had wolfed down a big metal alarm clock that had been left in a haymow. Reviewing the postmortem dissection I had performed earlier, I showed him the four stomach chambers and described the corresponding stages of rumination. When we came to the blocked pyloric valve—the same mechanism that allows food to pass from the stomach to the intestines in human digestive systems—he lost all doubt about the cause of death.

Pulling the deputy commander aside, Colonel Hackett told him that he had confirmed Perky's assertion, which the officers had dismissed as ridiculous on the trip down. The deputy commander turned to Perky.

"Ma'am, after your call I checked and found that there were in fact training exercises involving aluminum foil in this general area today," he said.

"Gee, Mister, for a minute there I thought I was imaginin' all this foolishness," Perky sniped. "Why didn't you tell me that to begin with?" The officer ventured nothing further.

We all proceeded to the barn, where I led an informal teaching round on the other sick cows. At the end of the tour, Perky turned to the deputy commander. "Well, Mister, what do you plan to do about this?"

"I'm sorry for your loss, but negotiating compensation is not my department, Mrs. Perkins," he said. "For that I'll have to send someone from our legal department down here tomorrow at oh-nine-hundred. I'll also send a cleanup crew first thing, and I promise you there won't be a scrap of aluminum left on your property."

Perky snorted. "I'll be waitin'," she said. "And let me tell ya right now, if this kind of thing happens again I'll take my shotgun to any of your airplanes that fly over my property."

The deputy commander's face registered no doubt. "Thank you for the warning, Ma'am," he said, signaling impending departure with a sharp glance at his entourage. They made hasty goodbyes, and the cars sped down the lonely road.

Perky and I stayed in the barn all night, administering a quart of mineral oil to each of the sick cows to help them pass the indigestible aluminum strips. Fortunately there were no more cases of bloat. At about 6:00 A.M. it seemed safe to leave the herd. Perky cooked us a big breakfast, and I returned to the office to start another day.

In the weeks and months that followed, I made many trips to Perky's farm, and gradually the sick cows regained their health and started producing milk again. She sued the government for the ill health of the cows and resulting loss of milk production, and I spent nearly as much time talking to government lawyers and writing medical reports as I did traveling to and from the farm and treating the cattle. The government attorneys refused to pay for the loss of milk production until I explained that in all animals, making milk is a luxury function of the body. As most farmers know, the healthier the mother, the more milk her body can produce. Conversely, illness, disease, or poor nutrition can cause milk production to shrink or dry up altogether.

Newly educated, the attorneys went back to the bargaining table with Perky, and it was a year before she agreed to their final offer. I never did find out what it was.

4

A Flight to Swans Island

In my early years as a vet, I worked with my older brother Phil in the practice he had established in Belfast. He owned a four-seat Cessna 172 airplane, and we both learned to fly at the local airport. This came in handy when we got the occasional calls from clients who lived on the many islands off Midcoast Maine. Flying saved countless hours, as the only alternative was a long drive and a slow ferry or a chartered boat.

One late afternoon in May, I received a phone call.

"Doc, that you? Freeland James, over here at Swans Island," said a familiar voice. "Got a draft horse that's cut himself real bad in the groin. He's bleeding pretty bad and I think I can see the end of a stick of wood or something sticking out of it. I've never seen horse gut, but I think there's some of that hangin' out there, too."

"Can you get close enough to slosh some disinfectant on the wound?" I asked.

"He's a corkin' good horse, but he's awful touchy about that cut. Won't let me near it. How soon can you come, Doc?"

The ferry slip to the island was about seventy miles away by land, which meant almost a three-hour round trip by car, plus an hour on the water each way. "What's the runway look like?" I asked. Freeland was a pilot himself.

"Nobody's used it this spring so far, she's been so soft and muddy. But I thought you'd ask, so I drove my pickup on it just before I called," he replied. "I think you could make it if you stay on the upper right side of her. Be no problem getting in. Getting out may be a little risky, but I'd chance it myself."

I knew that the time I'd save by flying could make the difference in saving the horse. Freeland confirmed that with his next remark: "He's piling up blood underneath him at a powerful rate."

"Okay, Freeland. Get some experienced horsemen out there and I'll be there in an hour," I said. "I have to get back here before dark, though, because we don't have any landing lights at the Belfast airstrip."

Leaving a half-dozen less urgent small-animal cases at the hospital, I drove the few minutes to the airport and was soon airborne, heading almost due east towards Swans Island. Landing on a twelve-hundred-foot muddy airstrip—the minimum length required for my plane—was dangerous, because the plane had what's called tricycle landing gear, a pair of wheels under the cabin and one under the nose. With the mud to slow me down, I didn't worry about going off the end of the runway, which led directly to the Atlantic Ocean. What did concern me was that the wheel under the nose could snub in the mud, causing the propeller to snag and shear off—or worse, send the plane tumbling end-over-end.

But before long I was circling over the strip, and I decided to risk it. As I set up my approach, I saw what Freeland meant about staying on the right side, which looked a lot firmer than the soggy ground to the left. But with a twenty-five-mile-an-hour cross wind coming off the ocean, my chances of staying on the right side were pretty slim.

I pulled the flap handle to slow her down and eased back on the throttle. The plane was still just about flying when she touched the earth and settled fast into a good six inches of mud. Shuddering, the aircraft plowed through the muck, the prop churning it up and throwing it all over the windshield faster than the wipers could keep up with it. I just prayed we would stop before the lip, because I couldn't do a thing. The normal rollout after landing would have been three hundred to four hundred feet, but the mire was so thick that we settled to a stop after only about eighty feet. I jumped out and saw that the entire craft was plastered with mud, but otherwise none the worse for wear.

Freeland came running over. "Gawd, Doc," he said. "Thought sure you'd lose your prop. That nose gear really scissored up."

"We'll look her over later," I said, "after we tend to the horse."

The horse, Max, was a high-strung eighteen-hundred-pounder, and he took an extra injection of tranquilizer before allowing me to get near enough to assess his wound. Even with Max under sedation, I was in a pretty precarious position kneeling under his hind legs to get a closer look.

Freeland's two horsemen turned out to be lobster fishermen, but they were game and very rugged. They tried to stay close to Max's head, holding his halter and the rope twitch I'd secured around his nose to divert his attention from me. What I saw told me immediately that I'd need to give him a general anesthetic, as a stick and a piece of bowel were indeed protruding from the bloody mess. After I injected three grams of intravenous sodium Pentothal, Max flailed for thirty seconds in delirium before settling quietly on his side, breathing deeply.

I moved fast. Tying ropes on all four legs, I enlisted Freeland and the lobstermen to roll Max on his back. After carefully pulling out a two-and-a-half-foot stick from the horse's abdomen, I washed up the protruding segment of bowel with saline solution, only to find it torn—a life-threatening condition. I stitched it up with about fifty "catgut" sutures. Catgut is an organic thread that is actually made of sheep intestine. I washed up the bowel again and replaced it, then reached into Max's abdomen as far as my plastic sleeve would allow, probing gently for further damage. To everyone's relief, I found no other injuries. After invading the abdominal cavity, which is extremely vulnerable to infection in horses, I sterilized it with three quarts of antibiotic solution poured through the wound, trimmed up the rough edges of the torn flesh, and sutured Max back together. I must have used yards of catgut.

Just as I was tying off the last suture, Max's powerful leg muscles started to twitch, a sign that he was coming out of anesthesia. I jumped up to help hold down his head, which is the only way to control a horse in field surgery. While anesthesia is wearing off, the brain, director of all bodily functions, often commands the body to get up before the legs are steady enough. This stage of entering or emerging from anesthesia is called delirium. The more nervous the animal, the more violent they tend to become at this stage, and Max was nothing if not nervous. His front legs and hooves flailed near our faces as we struggled to hold down his head.

We had some close calls but managed to escape being struck. Finally I saw that he was able to stand and we released him. I grabbed the base of his tail to steady him as he struggled to his feet.

The horse finally managed to stay upright, but wobbled and staggered about for fifteen minutes before his brain regained control over his body, at which point Freeland led him back to his stall. Not yet feeling the effects of his trauma, Max began gobbling the hay the men had placed in his crib. I gave him a tetanus shot and also injected antibiotics, leaving syringes, medication, and instructions for Freeland.

It was now nearing 5:30, and Freeland drove me to the airstrip. "You better get 'er up, Doc," he said, "elsewise you'll be puttin' down in the dark in Belfast."

We walked over to the plane and inspected it. Under the coating of dried mud, everything seemed to be intact. We tried to push it clear, but the wheels had settled deeply into the mud, and the plane was stuck fast. Hitching it to a nylon rope, Freeland managed to pull it onto higher ground with his pickup. We got a few pails of water from a house next to the airstrip and tossed them onto the windshield. Then we pointed her in the direction of take-off.

I suggested making a practice run to see if I could get airborne by the end of the short runway. But Freeland, a more experienced pilot, told me that a trial run would only chop up the already muddy field, making the final take-off even more precarious. Even though I was green, I had flown this particular plane about a hundred hours, and I knew its capabilities. I decided to give it a shot.

I was buckling my shoulder harness when I heard a tap on the window. There stood George Jackson, another islander. I opened the door.

"Hey, Doc, heard you were on the island. Mind giving my dogs their rabies shots? I'll drive you right over," he said.

"I've got to make it quick, George—I have to make it back home before dark," I said, explaining about the running lights.

"Jump in, Doc," he said by way of reply, opening the door of his 1939 Dodge pickup.

"Good luck, Doc," said Freeland, waving as he climbed into his own truck.

Three miles later, George pulled into his dooryard, which was lined with cars containing dogs and cats. "Having a party?" I asked. Obviously George or his wife, Elvira, had called around the island to announce an impromptu rabies clinic.

"Huh," he mumbled by way of reply, stroking his chin. I shot him a look, but he was gazing out his window, waving to an acquaintance. Fortunately or not, I had a large supply of rabies vaccine with me, as I always carried it for impromptu clinics. Had I not been trying to schedule a rabies clinic for the island for the past three months, I might have been tempted to tell George to turn around and drive me straight back to the airport.

Instead I moved swiftly from one vehicle to the next, giving rabies shots and dispensing free advice on medical problems ranging from rotting teeth to tumors of the rectum. One elderly lady's Pomeranian sank his sharp little incisors into my finger as I gave him his injection. "He's not gonna catch anything from you, is he?" she fretted.

I assured her I was healthy. Elvira Jackson handed me a molasses cookie and a Band-Aid as I climbed into George's truck. This little side trip had cost me a valuable thirty minutes.

"I'll collect the money for the shots and send it to ya tomorrow, Doc," he told me.

"Thanks," I said, "but meanwhile, we have to step on it, George."

He nodded with a look of grave purpose. I watched the speedometer needle climb from twenty to thirty-five miles per hour, and there it sat for the rest of the journey. When we pulled up to my plane, he turned to me with a proud smile. "That's faster'n I've driven in years," he said.

I was startled to see Freeland's pickup, a fire truck, and an ambulance parked next to the airstrip. Freeland was standing next to the vehicles with his lobstermen buddies and two other guys. "I got to thinkin'," he said, "maybe we should be ready in case something goes wrong."

"Thanks," I said, my courage dropping like a stone. As I moved toward the plane, one of the lobstermen caught up with me.

"That was quite an operation, Doc," he said.

"Thanks." I tossed the grips back into the passenger seat and climbed into the plane. The sun had now set.

"I got six boys, and the oldest one's pretty good with animals," the man continued. "Think he could ride around with you and learn to be a vet'nary?"

"Well, that's not quite the way it works," I told him. "Your boy's going to need six to eight years of college to become a veterinarian."

"Huh!" the man snorted, kicking at a stone with his boot. "Well, I reckon that ain't for him. He dropped outta school in sixth grade, just like his ol' man."

I thanked him for helping out with Max, then started the engine. I switched on the wing and cabin lights, ran the engine up to speed, checked the instruments, and pulled up the flaps. I waved to Freeland and gave him a thumbs-up, the sign that I was heading for take-off. He and the guys returned the signal.

With that I applied the brakes, pushed the throttle to full power, pulled back on the wheel to keep the nose out of the mud, and started to bump forward. I have never forgotten what happened next.

The plane wanted desperately to fly, but Mother Earth kept insisting that the Cessna would make a fine boat. In order to take off, this particular plane needed to reach fifty-five miles per hour. But jolting and jouncing along the muddy strip, I had all I could do to keep her headed in one direction, much less read the speedometer and other instruments. The seat of my pants was doing the flying. Three or four times my spirits lifted when all three wheels rode atop the mud, but these reprieves lasted only a few seconds before one or all the tires sank back into the mire, and we lurched and floundered toward the ocean.

Just when it seemed the runway would turn to quicksand, I felt the plane pick up speed—the last bit of the runway was ledge, part of the cliff that lay only a hundred yards ahead, and my wheels rode on solid ground. Now came the decision every pilot has to make at one time or another: go for it, or abort the take-off. I was heading straight for the cliff overhanging the rocky shore with the ocean dead ahead, but I gambled that the solid ground would allow me to gain enough speed to pull the wheel up sharply at the last second to get her airborne.

Gunning the engine, I yanked the wheel in tight to my stomach. The plane teetered forward. The stall warner, a sounding device that means the plane is about to fall earthward nose-first, blasted in my ears as the wing lights illuminated the car-sized boulders along the shoreline below. They seemed to grow larger by the second as the craft swanned over the steely chop of the Atlantic. I flashed on the film footage I had once seen of Doolittle's World War II raid on Tokyo, when pilots flew B-25 twin-engine bombers off a carrier. The planes lifted off the flight deck and settled dangerously close to the ocean as if to crash, but then pulled up at the last moment and began lumbering to a safe altitude.

That was how my little plane started to behave. The giant boulders grew smaller and smaller as the nose pointed skyward. The hull shuddered, shaking gobs of mud from its wheels. At some point the stall warner stopped squawking as I gained altitude, circled over the strip, gave Freeland and the boys a few flips of my wing lights, and headed due west. I was blessed with a clear night, and I was soon approaching Belfast along the coast.

I lined up for my landing, assisted only by the plane's landing lights on the dark airstrip. But as I started to settle in for the descent, it felt like a great hand was pushing me from behind, and the plane refused to touch down. We coasted past the airport windsock, and I saw that the wind had reversed direction since my departure. I gently reversed the flaps, poured on the gas, and circled around to line up for landing in the opposite direction.

Using the neon lights of the local ice-cream stand to line me up, I started my approach. Then, fifty feet over the woods at the end of the runway, a cloud of small winged bodies completely enveloped my windshield. In horror, I realized I was being swarmed by gypsy moths. I rolled down the side window, stuck my head out, and was instantly plastered with dozens of moths. I sputtered, shaking my head and blinking. Keeping my mouth closed and one eye open, I managed to put down on the grass next to the runway. After a few jarring lurches over the bumpy grass, the plane taxied to a stop.

The first thing I did was shut off the engine. As the prop slowed to a stop, I realized that my body, not the plane, was still shuddering. I sat until I stopped shaking, then half-fell out of the plane and kissed the ground. Maybe I should have kissed the Cessna instead.

True to his word, George mailed me the money from the impromptu rabies clinic before the week was out, and Freeland called twice in the next few weeks to update me on Max's progress.

"How's that plane of yours, Doc?" he asked me the first time.

"I'm giving her a rest," I told him.

"Hah!" he snorted, the closest thing to a laugh Freeland ever managed.

A month after my journey to the island, he called to tell me that Max had healed completely and returned to work with his old vigor.

5

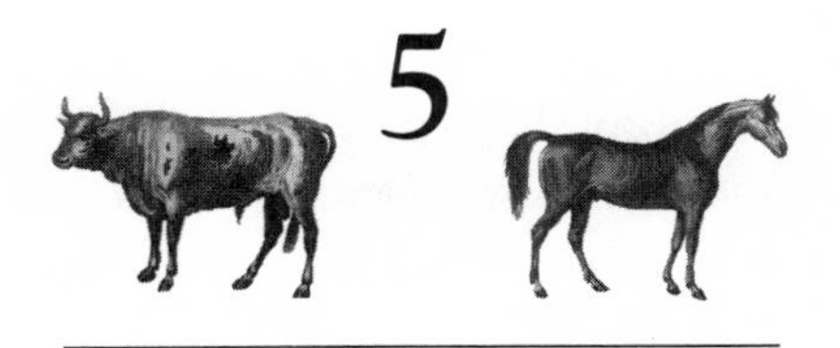

The Horse with the Broken Leg

One summer day I got a call from a lobsterman and regular client who lived seventy-some miles away. He was all out of breath.

"This is Maylon Crockett over in Stonington, Doctor Brown. You gotta get over here as quick as you can. My horse Blaze just fell and broke his leg. I was a riding him in the pasture and he was in full gallop when his front leg went into a woodchuck hole. His whole leg's just a danglin'."

"Pack it in ice, Maylon, and meet me at the airport. I should be there in about an hour," I said, throwing some things into my grip. "Are you getting a shower over that way?"

"Yup. But visibility is pretty good. You should make it in okay."

Putting aside less important work, I quickly left the office, loaded my portable X-ray machine into the car, and headed for the airport in a light, misting rain. By the time I circled the packed-gravel strip at Stonington, heavy rain beat against the plane's windshield in a steady torrent but presented no real problem as I touched down.

John Pushaw, Maylon's neighbor and fellow lobsterman, was waiting off the airstrip. He came over and helped me get my gear out of the plane.

"Call me Push," he said.

This took me by surprise. In those days most people in the rural areas along the coast—and especially lobstermen, an unusually taciturn breed—referred to themselves and other people more formally. Usually years went by before they would address you by your first name. We jumped into his car and headed toward Maylon's. "How's the fishin', Push?" I asked.

"Well, to tell you the truth, Doc, it ain't too good. Sea's been rough this summer and I've lost a pile of gear."

"Sorry to hear that."

"Ayuh, luck's been bad all around. My wife took off on me again. Told me she'd had enough of cuttin' lobster bait. So she run off with a young upstart from Deer Isle. Claims he buys his bait all cut. She's been off with him before. He gets drunk and pounds her round face almost square, then she comes wailin' home. Told her she could stay, but I weren't payin' a nickel to get her front teeth replaced. She'll have to pick crabmeat for 'em or go without."

Push paused to draw a breath. The shower was over and the sun shone again over a dramatically beautiful countryside of meadows meeting a shoreline of granite ledge. A thin fog veiled the ocean beyond.

"Maylon's horse looks pretty bad, Doc, worse than that million-dollar race horse that broke her leg awhile ago. They had to put her to sleep cause they couldn't handle her. Forgot her name. Been lots about it on TV lately."

"Ruffian, that's her name."

"Right, well, Maylon's horse looks to be a mess."

"We'll know soon enough." We turned down a dirt road, and I watched the ocean disappear behind a screen of evergreens.

"Hey, Doc, that machine you got there—that one of them X-ray machines?"

"Sure is," I answered.

"Well now, you couldn't take a picture of a person's leg with that, could ya?"

"Why not? They work the same on animals as they do on people."

"Well, I'll be damned," he allowed. "I supposed it'd be impossible to X-ray an animal."

"Why is that, Push?"

"'Cause they're shaped different, I guess."

"Actually, they're not," I told him. "Any animal with a backbone, including man, shares the same plan of anatomy throughout the five kinds of tissue that make up their bodies.

The only difference is that each kind of animal has developed in special ways to adapt to its environment. The same parts are there, just modified here and there."

"Like, I guess, a little foreign car and a Mack truck—they're both two kinds of the same machine, eh, Doc?"

"Exactly," I agreed. "They both have a motor, steering wheels, starter, tie rods, wheels, and all the other components needed to go on the road, even though they don't look alike."

"I'll be damned, Doc. Learn somethin' new every day," Push concluded as he turned into a driveway next to a pasture.

"Here we are, Doc, and there's your patient, out there in the field. Poor devil." Push pointed to a dark-brown Morgan crow-hopping on three legs, his right front leg held gingerly in front of him. Maylon was leading him gently forward by the halter. I returned his wave.

"By God, Maylon's got him almost up to the barn," Push exclaimed. "It happened about a half a mile from there. He's done pretty good to get him this far."

Push took out the X-ray machine and I grabbed two grips, and we walked into the pasture. Blaze was sweating profusely from pain as he hobbled to the lawn. Maylon's face was contorted with anxiety—he clearly hated to see Blaze suffering. I gave the horse a painkiller and took X-rays of his injured front leg while Push and Maylon held him still. The first showed a simple fracture of the tibia, or shinbone. A second X-ray revealed a dislocation of the medial patellar ligament of the knee—in other words, a dislocated knee. All four-legged animals, including horses, walk on the equivalent of a human being's fingers and toes. I explained this to Maylon and Push as they peered over my shoulder at the X-rays. Push looked puzzled.

"Animals have the same bones that we have," I continued. "Take your shoulder blade. Feel the same bone on Blaze here? It goes from the shoulder to the big upper-arm bone, called the humerus." Push nodded, fascinated. "Then from the elbow to the wrist, both horses and people have bones called the ulna and radius," I continued. "In your wrist, there are small bones called carpal bones, and Blaze has those, too, in what we think of as his wrists and ankles. But from there on down, Blaze's so-called foot is actually a modified hand. Feel those bones that form the top of your hand?" Push nodded. "A horse has the same bones. Those two that run from the top of your hand all the way to the ends of your middle and ring fingers form the front feet of a horse, with a giant fingernail wrapped around it, which we call the hoof."

Maylon could no longer conceal his anxiety and interrupted my anatomy lesson. "What are we gonna do, Doc? I hate to put him down—he's the best friend I've got. But I hate to see him suffer so."

"I suggest we do everything humanly possible before making that decision," I replied.

"But if he's got a hurt leg, how'll he get around while he's healin', Doc?"

"The same way he is right now, by crow-hopping. Just remember the good Lord gave him four legs to walk on instead of two. Take away one and he'll still manage to balance and move about. And as soon as he gets a little strength in that good front leg you'll be surprised how well he can walk on three legs."

"Boy, I always thought a horse with a broken leg was a goner, just like they talk about that poor ol' Ruffian," Push said.

"Not always, Push, especially nowadays. The same material and techniques used for repairing human fractures can be used on animals. In fact, most all new techniques and material for treating fractures are tried on animals before they're used on people," I said. "A veterinarian named Steinman developed and perfected the Steinman pin, which was widely used to set bones in both humans and animals during World War II and has been used ever since."

"Well, hell, then, why'd they put Ruffian down?" Maylon asked.

"To tell you the truth, I'm mystified about that myself, and a little shocked, too" I confessed. "They said something to the effect that they couldn't handle her, which to me seemed unreasonable, although getting only bits and pieces from TV and newspapers makes it difficult to make a judgment. But it would seem that with the same drugs and equipment available to both animal and human medicine, there wouldn't be a problem in setting the leg—especially with all the bone specialists involved in the case."

"Huh," Maylon grunted. "Sounds like maybe some insurance money mighta been involved. Do race horses have insurance, Doc?"

"You bet—millions of dollars' worth. Ruffian was worth millions as a brood mare, and her offspring potentially could bring millions also, so it's a distinct possibility that insurance came into play in the decision to put her down. Having not been involved, though, I'm only conjecturing."

Push whistled. "Well, that explains that," he said.

We all paused a moment, giving Ruffian her due before putting Blaze under anesthesia. After doing so, I carefully manipulated the joint, which snapped back into place and improved the leg's appearance immediately. Then, using a stainless-steel pin and lots of plaster of Paris, I aligned the fractured bone and created a cast that covered the leg from the hoof to just above the knee. Fortunately, it was summer, which would give my patient better conditions for healing during the months ahead. The front lawn of Maylon's house made a good operating table and provided an even surface for our three-legged friend to gain his footing when he came out of anesthesia.

After his operation, Blaze was groggy, but he soon struggled to his feet and made a few wobbly steps. Before long he was standing securely on three legs, and within half an hour he had mastered the feat of eating grass. I gave Maylon a crash course in using a hypodermic so he could help Blaze with the pain over the next few days. I collected up my gear and made ready for the trip back to the plane when he turned to me. "While you're here, Doc, my Uncle Newell wants you to take a look at his cat."

"How far away is he, Maylon?"

"About three miles. He ain't got no car. I'll drive ya down. It's on the way to the airport, anyway, if you've got time," he answered. "Ever since Aunt Wilda died Newell ain't set foot out of that cabin. They never did have any kids, and he's put all his attention on that cat now that Wilda's gone."

"We'll make time then," I said.

We said goodbye to Push and I thanked him for the ride. In the car I asked Maylon what his uncle did for a living.

"He's fished all his life. Makes traps in the winter and holds a town office. And of course, since Aunt Wilda went, he has to keep house," Maylon told me.

We arrived at Newell's small cabin to find nobody home.

"Huh," Maylon mused, looking concerned. "Generally Unc's back from lobsterin' by now."

"Is the cat in the house?" I asked.

"God, no, Doc. Bugs goes out in the boat with Uncle Newell every day. Goes everywhere with him, even to church. Unc claims he knows every hymn by heart, sings right along with the rest of the folks."

"What's wrong with Bugs?" I asked.

"That's quite a story," he replied, shaking his head. "He got himself in a mite of trouble in church last Sunday—attacked a lady sittin' a couple of pews in front of them. She was wearin' one of them little coats made of muskrat or some animal that had beady little eyes. Bugs leaped through the air and lit right into that lady's coat. Tore it to shreds before Newell could get ahold of him."

"Good lord," I said.

"That ain't the end of it. The lady beat Bugs with her Bible and he's been breathin' funny ever since. She's threatened to sue Uncle Newell. Unc told her she had no business wearin' dead animals around her neck, anyway. 'Sue all you want,' he says, but if she did, he said he was gonna sue her for bodily injury to Bugs."

"Did this all happen right in church?" I asked, wishing I'd been there.

"Oh, yes," Maylon said. "Hell of a scene, too. Reverend Rexford jumped down from the pulpit and tried to break 'em up. He allowed it weren't Christian of them to be fightin' in the Lord's house and that the Bible wasn't meant for beatin' up on God's creatures."

This hadn't been Bugs's only transgression in the House of the Lord. "Last Easter Sunday, Bugs leaped onto another lady's hat, but he didn't do much damage that time," Maylon reported. "Just tore off two little birdies on top and messed her hair a mite. Uncle Newell bought her a new birdie hat, and that was that."

After the most recent incident, Newell had carried the wheezing Bugs to a "vet'nary" nearby, Maylon continued, complaining that whatever medicine his uncle had given the cat hadn't done "a mite a good."

We walked down in front of the cabin, where Maylon led the way along a path through some shrubs. We emerged facing a ramshackle wooden staircase descending to the stony shore. Maylon shielded his eyes and stared out at the water. "There he is, Doc!" he shouted, pointing to a dinghy motoring to shore. "See, there's Bugs, settin' right on Unc's shoulder."

I watched as Maylon went to help Newell drag his dinghy onto the shore. Bugs had leaped to dry land as soon as the bow scraped rock. "You're late, Unc. Did you have some kinda trouble out there?"

"I'll say," Newell said. "It got blowin' durin' that shower comin' up through the reach, and I had to slow her back a mite." He eyed me suspiciously.

I stepped forward and introduced myself.

"For heaven's sake, May, why didn't you say who it 'twas? I mighta thought he was a revenuer and turned tail."

Maylon, a man in his forties, turned as surly as a boy of sixteen. "Revenuers don't wear coveralls," he growled, referring to the IRS. "And don't call me May."

Newell laughed him off. "Ain't that the truth. Last week a real revenuer was here for me," he said, shaking his head. "Them fellas ain't got nothin' else to do but chase some old lobsterman around for a few dollars. Hell, we must a passed a dozen half-million-dollar pleasure yachts out there today. I bet they get a lot more of a write-off than I do."

After a lengthy and repetitive history on Bugs's problem, we went into the house. Newell had let him in while we were talking. Now the fun began.

Almost always, examining an animal on its home ground is extremely difficult. In the presence of a stranger, the first thing it does is go to its favorite hiding place. Over the years I've had to scramble all over houses and barns in search of dogs and cats who didn't wish to make my acquaintance. Once I even had to tear out a wall, with the client's consent. Newell searched for Bugs under every bed and bureau in the two-story house, but the wily feline was nowhere to be found.

"Shouldn't a told him you might be a vet'nary, Doc," he said, scratching his head.

Finally, after an exhausting search, we found him pressed flat behind the refrigerator, wheezing like a busted accordion.

"Now Mr. Crockett," I said, trying to take charge of the situation, "Maylon and I will move the fridge out from the wall, and you grab Bugs. But you should put on some heavy gloves, because Bugs is pretty upset about strangers being around. When you get him, take him by the nape of the neck with one hand, and use your other hand to grip his hind legs. Stretch him out between your two hands, and he won't be able to bite or scratch you."

"Naw, I don't need no gloves," Newell assured me. "He won't hurt his daddy, will you, Bugsy?"

As we moved the fridge forward, Newell reached for the distressed cat. He was trying to get a grip on him when Bugs flew out of his arms and up the stairs, leaving a couple of ugly scratches on Newell's hands. "For God's sake, Doc. What's got into him? He's usually so tame."

"I'm a stranger, and I'm on his turf," I said. "Now we're all ganging up on him. If I were him I'd be on the lam, too."

Before we set off on the chase again, I repeated my instructions about the grip I had described earlier, as it would be the only way to hold him once we caught him again. A half-hour later it was my misfortune to peek under a small dresser and spot Bugs. I hadn't brought my leather gloves with me, and I just went for it. With a lot of determination and no small amount of blood loss, I finally managed to get him in my grasp. We all headed posthaste to the tiny bathroom and closed the door.

"Don't hurt him, Doc," Newell pleaded, as I handed Bugs over to Maylon, instructing him to hold the cat exactly the way I did. Maylon complied, and my stethoscope quickly told me Bugs had a collapsed lung, which meant he probably also had a ruptured diaphragm.

"Bugs'll have to go to my hospital for some surgery," I said. Newell's eyes were riveted on me as I explained that the tear in Bugs's diaphragm meant that intestines and other organs could be getting into his lung cavity. "All the lobes of his lungs are collapsed on the right side of his chest," I finished. "To leave him in his present state means almost certain death."

About a minute went by while Newell tried to clear his throat. His eyes filled with tears.

"I'm flying back right now and can take him with me," I offered.

"That will save you a trip."

After a couple of minutes Newell composed himself. "I don't want Bugs in one of those air machines," he said. "I'll bring him over in my boat first thing tomorrow morning. I'll get myself a room over there until he's ready to come home. He'll be all right, won't he?"

"He should be," I replied. "But it's a fairly serious operation, and there are no guarantees when it comes to blood and flesh. If he isn't repaired, though, he's surely going to suffer."

"He's all I got, Doc," Newell said miserably. "I'll be at your place with him at eight o'clock tomorrow mornin'."

"Okay, I'll put him on the schedule and do the best I possibly can to fix him," I told Newell.

"Oh, Doc, be sure you have lots of lettuce on hand," Newell told me as we walked to Maylon's car. "He loves it. That's how he got his name. His mother was a rabbit, I'm sure of it."

I didn't have time to give Newell a short course in genetics, so I assured him I'd lay in a supply of lettuce. Within the hour I was airborne, heading back to Belfast. The air was as smooth as silk and the sun shone on the island-dotted blue expanse below, a view that never failed to leave me breathless.

The next morning Newell was sitting with Bugs in a local taxi in my office parking lot when I returned from an emergency farm call. He asked if he could please stay in the room while I operated. Unlike most vets, I've never prohibited a client from watching surgery, but I've found that very few stomachs were strong enough to hang in for the whole procedure. Newell Crockett was an exception. He sat quietly on a chair staring straight ahead throughout the entire operation. I repaired the diaphragm, sealing the right side of his lung cavity. I gave him three or four small breaths mouth-to-mouth, to provide plenty of carbon dioxide, which stimulates the lungs to inhale fresh oxygen. Then I created the negative pressure needed to get him breathing on his own by gently squeezing the air from his lung cavity. Then I anxiously waited to see if the lungs would re-inflate. I gave a silent sigh of relief as they started to function once again by themselves.

During Bugs's recovery over the next three days, Newell stayed in a local motel and visited the cat several times a day. On the fourth day after surgery Newell ferried Bugs back home, where, I hoped, Newell would make as good a recovery as his feisty pet.

Six weeks later, Maylon borrowed a neighbor's horse trailer and returned with Blaze to take the cast off and remove the pin. The leg had apparently healed well, because Blaze was using it normally. After removing the cast, I saw that it had indeed mended well.

Newell had ridden along, holding Bugs on his lap, for a post-surgical checkup. As soon as Newell lifted him out of the cab, the cat got one whiff of me and headed for home. Luckily we caught up with him before he found either the highway or the woods. While Maylon kept an eye on Blaze, I took Bugs into the office and checked him over. His lungs were as clear as a bell, to Newell's immeasurable relief.

After that we went back outdoors to check Blaze. While I had been inside, Maylon had set a seaweed-covered bushel basket on the lawn near the horse corral. "Unc and I wanted to give you these lobsters we pulled this mornin'," he said.

I thanked them profusely—my family would have quite a feast that night. Newell gazed at Blaze, who was licking the skin of his leg where I'd removed the cast. "Well, how's the horse, Maylon?" he asked.

"Well, so far, looks like Blaze will be galloping again. Let's just hope there are no more woodchuck holes in his future. That one will never bother him again."

As the trailer wheels hit Route 1, I paused for a second or two and thought about the confidence I had gained from the experience with those two animals. Under certain circumstances it can boost one's confidence if you take a risk and win, which I had done with both of these animals. Considering the love and attention that the animals gave to their owners, and vice versa, I felt that the benefits had far outweighed the risks. And I happened to know, seeing them from time to time, that Bugs and Blaze both continued to lead normal lives and give their people a lot of happiness.

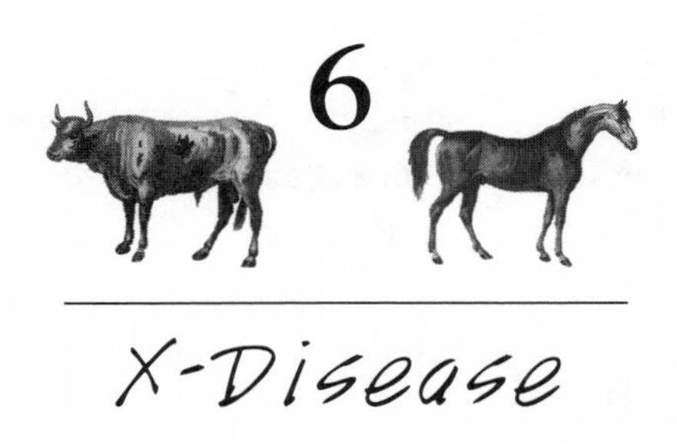

6

X-Disease

One late fall day I was called to the little town of Nobleboro, on the coast south of Rockland, by a small dairy farmer named Silas Toomey. He reported that he'd had five heifers die over the past four months. This morning he had discovered another dead cow, and several others in his remaining herd of fourteen looked sick. He implored me to come as soon as possible.

I had never been to Silas's farm before, so I spent a few minutes on the phone taking the history of the case. I learned that the cows had died since being let into one particular pasture. I made a mental note that this was also the time of year in Maine when pastures normally offer very little nutritious feed, a scarcity compounded this year by near-drought conditions over the summer. Before hanging up, I told the farmer not to bury the recently deceased heifer and that I would be there as soon as possible.

Several hours later I arrived at the property. As I got out of the car, Silas Toomey came over and introduced himself.

"I got to tell you right from the start," he said. "my vet'nary down here has been working on this case for four months now and my heifers just keep right on dyin'. And those that are still alive look like walkin' death. I'm afraid I'll have to get rid of the whole bunch of 'em, if they don't die on me first."

"Well," I said, "let's take a look." As we walked, Silas told me about the problem in more detail.

"The first cow took sick about four months ago. The sickness started just like the rest of 'em, with a dark, watery diarrhea. After that they just stop eatin', and their eyes sink into their heads. Pretty soon they're so weak they can't get up, and then they're dead in a week or ten days. My vet took a sample of some of the diarrhea and put it under his microscope. He saw worm eggs and thought they're probably dyin' of stomach worms, so he wormed the lot of 'em. He's done it three times since. But as you can see, there's been no improvement a'tall. In fact everything's goin' to hell in a hand basket."

I listened and sympathized with this hardship. Dairy farming is hard enough without mysterious epidemics, and I imagined he'd already spent considerable money on the other vet's services.

"Well," I said after thinking over what he told me, "in my opinion it's unlikely that you're dealing with an epidemic of stomach worms, because normally some of the cattle would have a natural immunity—they wouldn't all just up and die. Let's see what we can find out by doing a postmortem on this heifer that just died."

After a careful postmortem, the signs of disease remained ambiguous, partly because the corpse was now about six to eight hours old and the flesh had started to break down, and partly because the tissue changes and lesions I observed were typical of at least five diseases. I explained to Silas that to narrow down the possibilities, I'd need to take tissue samples and send them to a laboratory for further tests. He sighed, having heard this tune before.

After collecting tissue sections from the liver, kidney, bladder, and lungs, as well as a fecal sample, I re-examined the heifer's mouth. I was intrigued by the severe ulcers, averaging half an inch in diameter, in the dental pad in the roof of her mouth. (Because cows have no upper front teeth, nature has compensated by giving them a tough pad in the same place, allowing them to chomp like pros.) Another thing I noticed was how dehydrated the heifer was. The skin of the animal also attracted my attention. It seemed thicker than normal, in spite of the dehydration. And then a little flash bulb went off in my head.

Somewhere in my education I had heard of a condition called X-disease or hyperkeratosis. The condition is thought to be caused by chemicals called naphthalenes, which are present in some petroleum products. These chemicals are very toxic to animals (and people), causing radical changes in liver function, which in turn produce detrimental changes in other organs and systems of the body, including skin ulcers and thickening of the skin.

"Silas," I asked, "have you ever had any old cylinder oil in the pasture, or any other source of oil that the cattle might have gotten into in the last several months?"

He thought it over for a minute and said he couldn't recall anything like that. But as we were walking from the pasture to the milk room to wash up, a flashbulb popped in his brain, too. "Come to think of it, you know I turned out those critters about May fifteenth, and I had my manure pile out there that had to be moved like we do every spring," he said, growing excited. "I had two bucket tractors out in the pasture for four or five days while we was removing the manure pile, and those critters ate every sleeve off the hydraulic cylinders, and I wouldn't be surprised if they licked all the oil out of 'em, it being so dry this summer. On top of that, my boy was putting in some more fluid and he might have left the can out there. They could have lapped that up, too. Could that cause this?"

"If it's what I think it is, yes. Let's go on that assumption for now and see if we can keep these other cows alive."

"Sure would make me happy," Silas said.

"I have a lot of injectable vitamin A in my car. Why don't you see if you can get these critters from the pasture into the barn, those that can make it, and crowd them into a corner somewhere. I'll be back in a minute, and we'll inject them with some massive doses of vitamin A." I explained that this would bolster the integrity of the skin, which in this case was heavily affected as the poisoning progressed. "It won't cure the problem, but it will help strengthen their ability to heal," I added. As far as I recalled, there was no antidote for this condition, so my strategy was to prevent more harm and hope the cattle could begin to recover as the toxin worked its way out of their systems.

About an hour later I'd finished injecting his heifers. "As soon as you can, go buy a fifty-five-gallon drum of molasses and some 22-percent protein grain," I urged him. "Pour the molasses over their hay."

"What's that gonna do, Doc?" Silas asked.

"Hay doesn't have much protein, and extra protein will help their systems fight the disease. The molasses will give them energy."

I started packing up my grips, telling Silas I would call him immediately upon hearing from the pathology laboratory about the tissue analysis. I cautioned him that should my diagnosis be confirmed, the poison would take months to leave the cows' systems, and the milk should not be sold, nor should the heifers be bred.

"You know, I gotta say, you seem to be quite a good vet'nary," Silas said, walking me to the car. "While you're here would you mind taking a look at my wife's new foal? She's a four-month-old filly with a swelling on her navel about as big as a grapefruit. Gettin' kinda worried about it."

"Lead the way," I said.

"Well, the filly isn't here. She keeps it over to her sister's place. If you don't mind, we'll take a run over."

"Okay," I said. "Let's take my car, because it's got everything I need in it."

It turned out that Silas's sister-in-law lived twenty miles away in Tenants Harbor. Upon our arrival two women came into the yard to greet us. Silas introduced me to his wife, Pauline, and her sister, Thelma. When I asked to see the foal, Pauline told me she was in the pasture with her mother.

"She's not in a stall?" I asked.

"No," Pauline answered, "she's just staying down in my sister's pasture so she can roam a little bit."

"Well, why don't you fetch her as soon as you can, Pauline," I said. "I'm in kind of a hurry. This is an extra call for me today and I hadn't planned on it."

"That isn't going to be easy, 'cause I've never had a halter or a rope on her. She's a wild little thing."

"All right, let's get some grain and catch her mother, and perhaps the four of us can corner the filly so I can take a look."

We all set out to the pasture, and I told Pauline that foals should be handled immediately after birth and haltered within three days to avoid the situation that was confronting us. She took it in, but said it was too late with this little filly. The only reason they got her into the trailer to come down here was because she followed her mother. I had a sinking feeling I wasn't going to get a lot of other things done today.

The grain worked fine as bait for the mother. We caught her immediately. But our attempts to catch the filly were comical: we cajoled, we chased, we held still with grain in outstretched hands. She bested us in every maneuver. After about half an hour of these antics, we decided to stop fooling around and lead the mother into the barn. We knew the foal would follow and we could make our move in a more confined space.

We got them into the barn easily enough. Closing the door, we now had our patient contained in a 140-by-20-foot space. Before pursuing the filly, I told Silas what to do once I caught her. He nodded as if to say he'd just as soon avoid the whole situation, but none of us had a choice at this point, and he knew it.

A four-month-old foal, male or female, probably contains more power and energy, pound for pound, than any creature on earth. Catching this one was like trying to swipe a fly out of the air in a windstorm. Making matters worse, once trapped in the barn with us four humans, the mare knew something was up, and she became agitated, darting back and forth and snorting. Sensing danger, the foal stayed close to her mother, who did a commendable job of shielding the filly from our grasp.

Finally I made a lunge and literally tackled the filly. She struggled like a champ, pawing the air—and me—with flying hooves, slamming me with her head, and squirming like a four-legged snake. But I managed to hang on, and I reminded Silas about how to keep a good grip on her. When he had her under control, I gradually relinquished my hold on the foal and went about my examination.

The bunch on the foal's belly turned out to be a large umbilical hernia, which would require surgery. A hernia is a tear, or rupture, in a wall of muscle or membrane, allowing another organ or tissue to poke through the opening. The protruding organ or tissue can be squeezed so tightly that the blood supply is cut off, and the complications can be life threatening. When this occurs near the navel, it's called an umbilical hernia. This was the largest umbilical hernia I'd ever seen. It was half the size of a football and it contained small intestines that were in danger of being choked off, which would eventually kill the foal.

Turning to Pauline, I explained the situation and suggested that we do the surgery immediately. She gasped. "Are you sure we have to operate, doctor?"

"Definitely. Any rupture larger than three fingers wide isn't likely to heal naturally. And you could put two hands into this one."

She turned to Silas. "What do you think?"

"Well, I guess Doc knows what he's doing, Pauline. Why don't you try to calm down and listen to what he's saying," he said. "Doc, do you mean to do it right here?"

"Absolutely," I said. "We've got her in hand, and it needs to be done."

Pauline sighed and stared out the window. The mare, who had been pacing on one side of the barn, began tossing her head and snorting. She definitely didn't like us messing with her baby. The foal, picking up on Mom's mood, began to struggle again.

"Pauline, you'd better decide pretty quick," Silas warned, "This little filly is getting awful rambunctious."

After asking me more about the procedure and checking the cost, she finally consented to have it done, but said she couldn't bear to watch. She and her sister practically ran out of the barn. With an encouraging word to Silas I hurried out to my car to collect what I needed to perform the surgery. I decided to grab some tranquilizer for the mare, as she had the potential to seriously disrupt the operation. I was heading back when I heard a horrific crash of breaking glass.

I ran toward the barn. The big window was shattered, but I couldn't see what had caused it. I slipped inside the barn door to see the filly running crazily about with the agitated mare, whose face was covered in blood.

"The mare tried to jump through the window, Doc," Silas said. His voice was shaking and he was pale as a specter. "I think she put her eye out. There's blood everywhere. I couldn't hold the filly any longer."

"It's okay, Silas, just take a moment, take some deep breaths. I'll see if I can get a peek at that eye."

Talking as soothingly as I could given my own adrenaline flow, I walked beside the frantic horses until the mare allowed me to take her by the halter. Very slowly, I walked her over to a well-lit area and examined her wound. Her eye was intact, but the eyelid was hanging by a thread.

"It's okay, Silas," I said, "her eye is fine, but I'm afraid we're going to have to sew up a pretty bad cut on the eyelid."

Silas nodded, still looking pale and stunned. I slowly walked the mare over to him, talking calmly, as much for Silas as for the horse. "I'm going to need your help to hold her halter. You don't have to watch. Just keep talking to her, that will be a big help." He snapped out of his traumatized state, nodded, and took the halter.

I gave the mare an intravenous injection of tranquilizer, which immediately calmed her down. Then I trotted back to the car to get a bottle of local anesthesia. A shot of this near the wound blocked the nerves that give feeling to the eyelid. When I was sure she was numb, I proceeded to sew the eyelid back on. Silas held on like a trouper.

With that job done and the mare tranquil, I proceeded to catch the little filly again. Seeing her mother calm, the foal didn't give me quite such a hard time. Silas held her while I administered the anesthesia, and we laid her gently on the floor. I then repaired the rupture.

About an hour later the little filly was on her feet, albeit a bit wobbly. Pauline and her sister knocked on the barn door and entered just in time to see the foal walking about. Pauline flung her arms around her neck. Then she saw the mare's eye.

"What happened to my horse?" Her voice had an edge.

Silas explained, and Pauline burst into a rage. She accused us of being careless and incompetent, and she made it explicitly clear that she wasn't paying any vet bill for repairing the mare's eye. Silas tried to calm her down, but to no avail. I finally interceded and said that there wouldn't be any charge under the circumstances, but that I felt no neglect or fault for the accident. That seemed to calm her, and she fell silent.

Cleaning up my gear, I gave them careful post-surgical instructions for both the filly and the mare, as well as some medications for the mare's eye. Pauline, subdued if not contrite, listened intently.

Silas and I got in the car to head back to his farm. I started the engine. Just then I heard a hair-raising scream, followed by a soft thud. My head whipped around to see Pauline and her sister jumping up and down, shrieking and pointing to the back of my car. I shut the engine off and got out, only to find a bloody little kitten behind the car. Its hide was torn off the left side of its face and the skin hung off the entire left leg. But it was still alive. Cats, particularly kittens, often seek shelter under the hood of a vehicle, warming themselves on the engine, only to be killed or maimed by the fan blade when the engine starts.

"Well, little fella," I said, gently picking up the kitten, "at least you picked the right car." I asked Pauline to get a clean towel, and when she returned, placed it on the hood. Laying the kitten on it, I gave it anesthesia and proceeded to stitch its skin back on. Once more I left medication and instructions before heading out for the road.

Having been delayed another half an hour by this impromptu surgery, Silas and I headed back to his farm.

"Do you run into this sort of thing every day?" he asked.

"No," I said. "This has been one of my better days."

Over the next few weeks when I had calls in Silas's area I would stop in to see how my patients were progressing. The heifers, except for one more that died the day of my visit, all recovered. Though I felt bad to lose that one, I realized that Silas was fortunate to have saved the rest. Both horses healed just fine.

Several months later Silas and Pauline sent me a check for my services, including the mare's eyelid surgery. Pauline even enclosed a note apologizing for the way she had acted. "You can't even see the scars on the foal or the mare," she wrote.

The following Christmas I got another surprise. A card arrived from Pauline's sister. When I opened it, a color photograph fell out. It was a picture of a handsome black and white cat sitting on top of a piano. The card read: "Dear Dr. Brown, thanks for saving my life. I don't go near cars anymore. Your grateful friend, Fanblade."

7

Watson

One crisp fall afternoon a farmer named Perley Russell called from Penobscot, about forty-five miles from my office, to ask me to treat a case of bloat in three yearling heifers. Bloat, a potentially fatal condition, constituted an emergency. Over the phone I told the farmer to give each heifer an ounce of turpentine in a quart of milk and then gag her. The gag, a rag looping behind the back molars and tied around the head, would keep them chewing and belching up the gas until I could get there.

At about two o'clock I arrived at a shabby little farm, which I had never before visited. The falling-down house and dilapidated barn looked like the Dust Bowl photographs from the Great Depression. Beside the barn was a small cleared area covered with ground juniper, small pines, and fir trees. Fenders, bumpers, and all manner of debris from ancient vehicles were strewn about.

After introducing myself, I accompanied Perley Russell into the rickety barn. Inside were three of the scrawniest heifers I'd seen in quite a spell. Being so thin and so bloated, they looked as if they were about to float into the air like balloons. There are many causes of bloat, a rapid overproduction of lethal methane gas in the second stomach compartment, the paunch or the rumen. One of the most common causes is a change in feed, particularly when animals switch pastures during the grazing season. The biggest offenders are clover and other legumes, which are lush and heavy at their peak growth. But obviously Perley didn't have any such valuable feed around his place: I hadn't seen so much as a square foot of grass on the premises. Genuinely curious, I started to quiz Perley about what the heifers were getting to eat.

"What you see out there in the pasture, Doc," he answered.

"Pasture?"

"Yeah, that little yard out front."

With a shock, I realized he meant the metal-strewn lot crowded with weeds and evergreens.

"Plus," he continued, "I'm buyin' a little chicken grain that gets spilled around the box cars down at the railroad yard. Then once or twice a week I go over to the bakery in the village and buy day-old bread. Fella down the road told me bread was good for a cow. I get it for practically nothin'."

While he talked I went about treating the cows. First I numbed an area on their sides with local anesthesia. Then I used a trocar, inserted in a metal tube, to punch a nickel-sized hole through the skin and into the paunch. I then withdrew the trocar, releasing a whoosh of gas. As I worked I continued to ply Perley with questions: when was the last time he had given them bread? How much had they eaten? He told me he'd fed them bread that very morning, having stopped by the bakery on his way home from his job at the mill.

As time passed, it became obvious that the standard treatment was not keeping pace with the production of gas. I went to the car and got a surgery kit and after injecting a local anesthetic to numb the tissue, I made a twelve-inch incision in one of the emaciated heifers and carefully sliced through the skin and three layers of muscle to the rumen. I withdrew the pouchy organ and incised it. One look at the contents told me why treatment had been ineffective: it was stuffed full of raw bread dough and yeast.

I turned to Perley. "Have you ever given her this uncooked bread dough before?"

"Nope, this is the first time," he said. "I got there kinda late and they'd sold all the day-old bread, so the fella asked me if I wanted the leftover dough. I s'posed it'd be all right, so I fetched up about fifty pounds of it. Divided it up amongst the three of them. Never thought it would cause trouble."

"Well, if it were just plain dough, you'd probably be right. But there must be a pound of straight yeast in that mess, plus the yeast in the dough."

The farmer looked stricken. "You think you might be able to save 'em, Doc?"

"Difficult to say, Perley. Let's get these other two opened up and clean that dough out of them. Then we'll see what happens."

With that we set to work. The surgeries consumed about two hours and went pretty well, considering the animals' poor overall physical condition. As we were finishing the last one, Perley's wife came into the barn.

"Perley, don't forget the monkey's tooth," she said, nodding curtly to me and stalking back to the house.

"I won't, dear," he said to her retreating figure.

"What monkey's tooth, Perley?"

He sighed. "Well, my wife has this pet monkey, and lately he keeps chewing to one side. His gums are swollen up and he's gettin' kinda fussy about his mouth. Though truth be told, he's always been a biter."

I knew where this was heading even before his next sentence. "While you're here, she'd like to have you take a look at him if you would."

"Okay," I agreed. "As soon as we finish up here."

"That's mighty kind of ya, Doc," he said. "Wife sure would appreciate it. I ain't too much on the monkey, to tell you the truth. Can't trust the little bastard. Christ, just when you think he's gettin' friendly, he takes a crack at ya."

"Gee, he sounds irresistible, Perley," I said.

The humor passed right over his head. "The wife sits with him all day long, pickin' off fleas and scratchin' his back," he rattled on, "but once in a while he'll even give her a pretty good rakin' over with them sharp little teeth. I suggest you put on a pair of gloves before you go near that character, Doc."

"You don't have to tell me twice," I said. I had treated the occasional monkey in my small-animal hospital over the years, and most of them fit Perley's description.

I stitched the incision on the last cow. "Hey, Doc," Perley asked, "seein's how there's so much of that bread dough you cleaned out of them, do you s'pose it'd be all right for the pigs?"

"Nope," I said. "I don't believe that you should feed it to man nor beast. Just discard it and think of this as an educational experience."

Then I asked Perley if he had any neighbors with cows.

He allowed that Weldon Pickett had a few, down at the other end of town. "Why do you ask, Doc?"

"Well, in order to start those rumens again it would be wise for us to get a little cud material out of some other cows. It would act as a starter for these little heifers."

He scrunched up his nose. "What's a cud?"

"After a cow chews and swallows her food, the chewed food, called the cud, comes back up into her mouth and she chews it again. This aids the digestive process, which is called rumination," I explained. "What we have to do is catch a few of those cows when they're chewing their cuds and reach down into their mouths, pull out some of the cud, and bring it back here. Then we'll take a balling gun—that's an instrument for putting large pills down a cow's throat—and put some of the cud into each of your cows' stomachs. It will start up their digestion again."

Throughout my description, Perley looked increasingly uncomfortable, shifting his weight from one foot to another. "Doc, I hate to sound chicken, but I ain't about to put my hand down a cow's throat. I've seen them jaw teeth on 'em. They look like a mess of mountains with sharp peaks all over."

"Don't worry," I said, "I'll be doing the cud retrieval. I'm used to sticking my hand down large animals' throats—we had to do it in front of a professor before we could graduate from vet school."

"You're kiddin' me," he exclaimed.

"Nope," I said. "In fact it will be a damned sight safer than working on that monkey's mouth."

"You got that right," Perley allowed.

I asked him to go into the house and phone his neighbor and ask if we could take the cuds out of a few of his cows.

"I ain't got a phone. I always use my neighbor's. Probably just as quick to run up there."

After a bumpy trip up a back road we arrived at Weldon Pickett's farm. Perley knocked on the door and out came Pickett. After explaining the plight of Perley's cows, I saw he was reluctant to go along with my plan. "Very seldom have I had a cow live after she lost her cud," Pickett said.

I explained that when a cow gets sick for any reason, the first thing that stops is rumination, and losing the cud was a sign of illness, not the cause of it. Hearing me out, he agreed to let us proceed. Half an hour later I had retrieved about a quart of cud material from the mouths of several cows as Perley watched in awe. Finally we had enough, and we thanked Pickett and returned to Perley's place.

Once there, I filled some one-and-a-half-ounce gelatin capsules with cud material, inserting the balling gun into the cows' mouths, and injected the big homemade pills into their stomachs. "Now," I said when we were finished, "let me tell you something, Perley. I suggest you don't feed those critters any bread dough again. Just be sure that it's been baked into real bread, and don't accept any yeast either. It's not the best feed, but if you need to supplement their diet with it once in awhile, that won't hurt. But these animals need more than that. They're awfully thin, and they need to get some nutritious hay on a regular basis, preferably alfalfa, but clover or timothy will do as well. Otherwise they'll get sick from malnutrition. Are you hearing me?"

"I hear ya, Doc."

"Okay," I said. The sun was setting as he accompanied me to my car, where I pulled out my leather gloves. "Now, let's have a look at that monkey."

He led me into the house and introduced me to his wife, Una. The monkey, Watson, started chattering from his perch atop the refrigerator the moment we walked in. He was a big son of a gun, a rhesus monkey about the size of a cocker spaniel. "Wat-son," Una sang, and he hopped into her arms. But when I reached out for him, she yanked him back.

"Promise you won't hurt him," she commanded.

I assured her I wouldn't, but I also warned that he might not like having me pry his sore mouth open and would probably protest. She sighed and nodded.

I approached Watson slowly and started stroking him on the head with my gloved hand. He squawked but didn't move. Talking in a soothing voice, I gently placed my hand on his lower jaw to have a look at his teeth. I saw them only for an instant, but I sure felt them. He sank all four canines right through the glove and into the flesh of my hand, drawing blood. Then he flashed to the top of the refrigerator, where he shrieked like he was the one who'd been bitten. It all happened so fast I barely saw him move.

"Okay," I said, "the only way I'm going to be able to work on his mouth is to give him a general anesthetic."

Una refused at first, but after I explained that I'd use a safe, short-acting drug, she supposed it would be all right. I dug a syringe out of my grip and withdrew a light dose of anesthesia. Then I enlisted a grim-faced Perley to don my gloves and call Watson down from the refrigerator.

"Get a good grip on him for about five seconds," I said in an encouraging voice. "That's as long as it will take to give him a shot."

Grumbling, Perley moved toward the refrigerator, calling Watson in an unnaturally sweet voice. Una darted through a curtain into an adjoining room. "You best not harm one hair on Watson's head!" she screamed.

Well, I thought, that ought to calm old Watson down. Perley continued his approach, clucking and calling. When he raised his arms to seize the monkey, Watson sailed over Perley's head and grabbed the hanging light fixture in the middle of the kitchen ceiling. There he swung, chattering a shrill warning.

Perley lunged for the light fixture, but Watson was ahead of him again, leaping to the drain board by the kitchen sink, where he started chucking forks and spoons and knives and everything his little hands could find in our direction—with a fair degree of accuracy, I might add.

"Why, you little cuss," Perley muttered, making another dive for the enraged monkey. Watson again evaded his grasp and flew through the curtain into the adjoining room. He jumped into Una's arms and was hugging her neck for dear life when Perley and I entered the room. "Perley, here's our chance," I said in a low, measured voice. "Just take him off her shoulder. Una, just stay calm and let Perley take him from you."

As Perley reached for him, the little fellow screeched like a banshee, but Perley steeled himself and managed to peel Watson's fingers from Una's neck. With the monkey finally in his grasp, Perley started to turn toward me. Then, with inspired grace, Watson twisted his body around and urinated all over Perley's face. Cursing, Perley let go, liberating Watson once again.

Unbeknownst to us, Una had left the bedroom window open about four inches, propped up by a block of wood. Even if I'd known about it, I would have bet that a monkey the size of Watson couldn't fit through that opening. And I would have been dead wrong.

After his brilliant urination maneuver, Watson darted into the bedroom and wormed his way through the window on his way to freedom. And he would have made it, except that he had dislodged the block of wood in his exit, and the window had fallen on the end of his long tail. He hung upside down outside the window, flailing and squawking, with about six inches of his tail still inside the bedroom. By this time Perley needed no instruction. He grabbed the tail as I hoisted the window open and flung Watson back into the room, sustaining several hefty bites.

Una flung herself facedown on the bed and began sobbing into the pillow. Perley ignored her, having finally achieved a solid grip on Watson's neck. We proceeded into the kitchen. Watson's adrenaline was peaking now: as he thrashed and struggled, he sprayed feces all over Perley, me, and the kitchen.

"Hang on, Perley!" I cried, pulling my syringe from my shirt pocket. After three or four tries I managed to secure one of Watson's flailing hind legs. No sooner had I inserted the needle into his taut muscle than, quick as a flash, Watson reached down, plucked out the syringe, and jabbed it into my forearm. Perley, now exhausted, relaxed his grip, allowing Watson to free himself.

What happened next I would prefer to forget. During a tense moment when we were trying to subdue the monkey, Perley's neighbor, Joe, entered the kitchen unnoticed. After Watson sank the needle into my arm, he streaked behind Joe, who was cowering near the kitchen door.

"I, I guess I'll stop by another time," murmured Joe, who had one hand already on the doorknob. He pulled the door open, and Watson fled into the starless night.

"Grab him!" shouted Perley, dashing out the door.

His shout brought Una off her bed. She stormed into the kitchen and came right up to me, the other two having gone off in pursuit.

"He's never been outdoors except on his leash, and now he's out there, we're never, ever gonna see him again!" she hissed, jabbing her finger in my face. "I'm gonna get ahold of a lawyer first thing in the morning and start legal action against you, Brown. You can bet your bottom dollar on that."

At this point neighbor Joe came charging back into the kitchen. "I'm sorry, Una, but I don't dare to be out there with that ugly beast roamin' about," he gasped.

Perley followed close behind. Still splattered with monkey urine and feces, he went to the sink and washed off his face and his hands, cussing quietly. I examined the puncture in my arm. Watson hadn't had a chance to push the plunger once he'd nailed me with the syringe, so except for the soreness I was in relatively good shape.

Una ran outside, and we heard her calling for the monkey. Perley, Joe, and I sat down at the kitchen table to map out our next move. Joe suggested he go home and get his 30-30 rifle, "just in case we have to shoot the little varmint."

"Hush!" Perley yelped.

Una was just coming back into the kitchen as Joe uttered those words. "I'll put three bullets into you for every one you put into Watson!" she allowed. "You can bet your bottom dollar on that." With that she stalked into the bedroom, returning moments later with a loaded rifle. "Just in case you don't think I know how to use this gun, I wanna show you somethin'," she snarled.

"Now look, Una," Perley said. "I know you know how to use it, so why don't you put that away before you hurt somebody."

She whirled to face him. "You sit down and shut your mouth, Perley Russell. You see that Chapstick on that windowsill over there? Just watch it."

She fired. The shot not only blew the Chapstick to bits, it blasted the windowsill and a pane of glass to smithereens. The chill night air streamed in around our shoulders.

"There," she said. "You got any more ideas about shootin' my monkey?"

Without another word, the three of us men grabbed flashlights and went out into the blackness to look for Watson. After an hour's search of the small yard and surrounding woods we saw no sign of him and reconnoitered in the kitchen. Perley asked Joe to run home and phone some more neighbors to come over and help. Joe acquiesced, but seconds later he burst back in, breathless and wide-eyed.

"Perley," he cried, "He's under the seat of my pickup."

Oh, boy, I thought to myself, having seen many an animal lash out from under a car seat. Joe had parked in front of the porch, leaving his driver's-side window half-open. Apparently Watson had decided that the truck was a good place to hole up.

"My poor baby!" Una shouted.

Grabbing one of the flashlights off the table, she announced that she'd take over from here. Joe, Perley, and I filed out onto the porch to observe. She approached the pickup, cooing, "Come on, baby. Come on back to Momma, come to Momma. Yeeeesss, these people have treated you so badly. Now you can come to Momma."

Perley shook his head, a mannerism I now recognized as a habitual gesture of hopelessness. Una opened the cab door and squatted to peer under the seat. "Come on, baby. It's all right now," she cooed, extending her hand slowly toward the dark cavern.

A moment later she let out a scream that was probably heard in three counties. In the dim light of the porch bulb, we saw her recoil from the pickup as if she'd been shot. She landed on her posterior in front of us, clutching her right hand. Perley helped her up.

She retreated into the kitchen, and we followed. I stood beside her as she examined the bite wound in her hand. It was deep and bleeding copiously.

"You need to clean this," I advised. "Do you have any hydrogen peroxide?" She nodded. All the fight had gone out of her. "It's not going to feel good," I continued, "but you should flush out the wound with cold water and then pour the hydrogen peroxide directly into the punctures, the more the better. After you do that, I'll bandage it for you." I stressed that she visit her doctor for a tetanus shot and more treatment.

Meanwhile, Joe was complaining that he couldn't go home until we got that monkey out of his truck. We agreed that I would take a turn again. Donning the gloves, I went to the truck with Perley, who took along the porch broom. Opening the driver's-side door a few inches, I bent down and shined my light under the seat. The beam picked out the two glittering dots of Watson's eyes. He hissed at me. I nodded to Perley, who took my place at the door, while I went around to the passenger-side window, shining my light into the cab. Perley slid the broomstick gently toward the monkey. Watson greeted it with a series of blood-curdling screeches. Then he attacked the pole with full fury. So intense was his assault that he failed to notice Perley slowly pulling the broomstick, dragging him out into the open.

At that moment I flung open the passenger door and grabbed Watson by the midsection while he clawed at my gloved hands. Holding the squirming monkey at arm's length, I scooted into the kitchen with Perley right behind. Joe lingered on the porch. At this point, Watson wasn't the only one with adrenaline pumping through his veins. I held him up to the kitchen light and squinted into his wide-open mouth as he howled and screeched.

"By God, Perley, he's missing one of his canine teeth," I said. "It must have been the sore one, because the gum's all swollen and irritated around the socket." I released Watson, who tore off, screaming, to Momma's bedroom.

After the dust settled, Joe, Perley, and I searched the pickup for the missing tooth. No luck. I retraced my steps from the yard into the house, and there on the porch floor was the tooth. "Gee whiz, Doc," said Perley. "Guess he must've loosened it up on the broom handle. Nasty lookin' thing, ain't it?"

"In more ways than one," I agreed. The eyetooth resembled a tiny, rusted scimitar. "Looks like his sweet tooth. You might suggest that Una cut back on his sugary snacks."

"Hah!" Perley snorted. "You tell her that."

Joe departed, and Perley and I walked to the barn to check on the heifers. The healing process was just beginning, but with their distress alleviated, the young cows looked livelier than they did when I arrived. As I was about to leave, Perley started shuffling his feet. "About your bill, Doc," he said. "I know I owe you quite a bit of money, but my next check's goin' all to payments on the house and the TV. If you give me a month I'll try to even up with ya."

"Well, what choice do I have, Perley? Do the best you can." With that I hightailed it back to Belfast, having spent almost five hours at the farm.

About a month later Perley called me. "The reason I ain't sent you no money is that we had to pay another vet'nary to operate on Una's monkey, 'cause he got some kinda bowel problem. You remember Watson, don't ya?"

"I vaguely remember Watson," I said, glancing at the scars on my hands.

"Hah! I reckon so," he said. "Anyway, don't worry, Doc. I'm gonna send you somethin' next week."

He did send something the next week—a dollar. And that was all that ever came. After six months I told the bookkeeper to stop billing him.

8 Blackwater

At about three o'clock one February morning, my bedside phone rang. The caller was Duane Clawson, a logger in the coastal village of Blue Hill about fifty miles away. In an anxious voice, he told me that one of his logging horses had a bad case of colic. Duane was one of the last people in the area to use draft horses to haul or "twitch" logs out of the woods.

"Can you come, Doc?" he asked. "He's rolling and thrashing in his hovel, staving himself up pretty good. The place is gonna be matchsticks by morning."

While listening, I had stolen out of bed to have a look at the weather, only to see wind-driven snow falling at an alarming rate. "Sounds pretty serious, Duane," I said. "I'll come as soon as possible, but I don't like the looks of this storm. What's it doing out your way?"

"Pretty windy, but the snow is just startin' up."

"Well, I'll get going now and hope for the best," I said, pulling on long underwear.

"Thanks, Doc," Duane said. "I'll meet you at the general store in the village and take you the rest of the way in. He's up in the woods about three miles."

I set out in a swarm of whirling flakes. Wind blasted the snow across the road in thirty-mile-per-hour gusts that rocked my car. I had pretty good snow tires, but my car fishtailed wildly every few miles as the white stuff piled up. By the time I arrived at the general store in Blue Hill, six inches of snow covered the road, and it was nearly four-thirty. The place wasn't open yet, but Duane's beat-up pickup was parked out front, engine running and headlights glowing through the curtain of snow.

"Thank God," he said. "You made it. Follow me."

I trailed him out of town about three miles, where we turned off onto a gravel road—not that any gravel was visible at this point, but I knew it from previous calls. The storm had continued to gain fury as we continued for about a mile and a half, where an opening in the roadside trees was barely visible. Duane parked his truck and climbed out, holding a dim flashlight in one hand and shielding his face from the wind with the other. I rolled down my window just as the local radio station reported that the wind chill factor would reach thirty-five below zero by dawn.

"This is where we get out, Doc," he said, shining the light on the space through the trees. "He's up that woods road about a mile or so. The vehicles won't make it in the snow, so we're gonna have to walk."

I climbed out of the car and tugged the hood of my sweatshirt over my head as I popped the trunk. The wind howled like a hundred wolves. I dropped all the drugs and syringes I thought I'd need into my steel pail and poured in some disinfectant solution. Then I grabbed my grip and an extra flashlight and followed Duane into the woods.

The temperature was plummeting, turning the falling snow to the consistency of Styrofoam. Every step crunched as if someone was chewing crackers in my ear. It seemed like we tramped forever before arriving at Duane's logging camp—a crude cabin for him and his teamster, and a hovel for his horses. Even over the keening wind I could hear the horse's throes of agony, which sounded like a shipwreck in a gale. The 2,000-pound beast was writhing and crashing against the walls of the twelve-foot-square shelter.

Inside, Duane's teamster, Lionel, stood next to the door holding a kerosene lantern. The horse was rolling on its back, and the worn caulks on his winter shoes reflected in the lantern light. Moaning and twisting in pain, the miserable creature slammed his body from one wall to another. Only a tough nylon rope tied to his halter and made fast to a hook in the ceiling prevented him from making sawdust out of the building and hamburger out of us. People who work with horses know that the only way to control a horse, if at all, is by the head, where it is most sensitive. Horses have a low pain threshold, lower than most other vertebrates, including humans.

I spent a moment or two studying the poor beast.

"What d'ya think's wrong with him, Doc?" Duane asked, his voice anguished at the sight of his daily companion suffering so mightily.

Some animal owners seem to think that a veterinarian possesses mystical diagnostic powers, as though he or she can look right through the skin of an animal and read a cue card on which the diagnosis is written. I wish I had that ability, but in fact I have to go through a mental checklist accumulated through training and years of experience. One thing was certain, this was no harmless case of colic.

As a diagnostic term "colic" is virtually meaningless, indicating only pain in the abdomen. This symptom can result from any number of underlying conditions, and in the early stages, it can be nearly impossible to determine what's causing the pain. It could be a simple case of gas, with no lasting consequences, or a fatal blockage, twist, or tear in the intestine. In some cases, a horse will gorge itself on grain or pulp that swells with moisture. As the grain expands in the stomach, it causes agonizing pain and can even burst the stomach walls. Because horses' digestive systems don't allow them to rid themselves of the excess bulk by vomiting, this condition—called bloat—demands emergency intervention. All these possibilities were flashing through my brain as I contemplated Duane's workhorse. But as yet, I had no idea what was happening under the skin.

"I don't know yet, Duane," I answered.

My first task was to reach the horse's head and try to inject a painkiller into a vein. Like most horse hovels, this one had only one entrance, forcing me to risk being pinned in the far corner with no escape as I ventured toward the animal's head. I sucked in my fear and loaded a syringe with a mild dose of anesthesia and tranquilizer. After studying the rhythm of his violent motions, I saw my opportunity and darted toward his head. But I was not quite fast enough.

In a flash the horse rolled, trapping me under him. I felt like someone had stuck a knife between my ribs as I was crushed against the floor. The poor brute must have felt more comfortable with my body under him to break the pattern of his excruciating pain, because he paused there for a few moments. I tried to choke back my panic as my vision blacked out from lack of oxygen. Like so many times before and since, I thought I'd really "bought the farm." Then the horse rolled off momentarily, resuming his rhythmic thrashing. I came to in time to hear Duane tell Lionel to fetch a rifle and shoot the horse. "Quick," he shouted, "or Doc's a goner!"

My head clearing fast with the help of adrenaline, I scrambled to my feet seconds after the horse rolled away from me. "Wait!" I shouted to Duane. I lunged toward the horse's head, where I saw my glass syringe on the floor, miraculously still in one piece. I grabbed it, dodging the flailing hooves, which could shatter my skull like a pumpkin.

Duane turned to face me, amazed.

"Duane," I shouted, "toss me a big cotton swab with alcohol on it."

With Lionel holding the lantern, Duane rushed to oblige. My college baseball days paid off as I fielded his toss. Then we got a real reprieve: for a few seconds exhaustion overtook my patient, and he lay on his side huffing like a steam engine. My fingers found his jugular vein, and I lost no time squeezing the contents of my syringe into it. The beast relaxed instantly, his breath labored but his body quiet at last. I could now observe the flecks of foam on his coat and feel his galloping pulse. These symptoms could have been a result of his contortions, but they also could have been present when the pain began. I inserted a rectal thermometer and found that the horse was running a high fever. I ran my hands over his body and felt knotty muscle spasms in his hindquarters and rear legs.

"Lionel, did you notice anything unusual before he started thrashing?" I asked. Because of the storm, Duane and I hadn't had a chance to talk about the early symptoms.

"I put his feed down last night and he seemed fine," Lionel replied.

Duane concurred. "We worked 'im pretty good last week, so we've been resting 'im over the weekend." He shrugged. "Next thing I know, I hear this crashing like to wake the dead. We come out of the cabin and found him just like you saw."

Monday morning—right on time, I thought to myself. What we were looking at was a serious case of azoturia, also known as myohemoglobinuria and a few other scientific names. Farmers tend to call it "Monday morning sickness," "tying-up sickness," or "blackwater," after the color of the suffering animal's urine.

This condition usually manifests itself after a horse works hard for five or six days and then stands idle over the weekend or during a patch of bad weather. Technically, any vertebrate can get this condition, but workhorses are the most vulnerable because they have such large quadraceps and are used for heavy labor. Usually the farmer or teamster doesn't notice the problem until it's time to hitch up the horse again on Monday morning.

The problem arises when the horse consumes a working ration of feed while stabled. Because the body is suddenly inactive, the liver and muscles get overwhelmed by the big load of carbohydrates and can't properly metabolize them. Glycogen, a product of carbohydrate metabolism, accumulates in the muscles and turns into lactic acid instead of breaking down all the way to carbon dioxide and water. This short-circuited metabolic cycle causes a gamut of problems, including painful cramping in the large muscles in the hindquarters and even the breakdown of the muscle tissue, which can result in permanent damage and lameness.

In advanced cases the damaged muscle releases myoglobin, a muscle protein, into the bloodstream, where it travels to the kidneys and then into the urine. The dark-red protein turns the urine the color of coffee perked on an old woods-camp stove—hence the name blackwater. But often, as in the case of Duane's horse, the animal doesn't pass any urine during the acute phase of the disease, making it easy to confuse with other causes of colic.

As wind-whipped snow sifted through the seams of the crudely built shelter, I sent Lionel to fetch a dry towel to rub down the sweat-soaked animal and a blanket to keep him warm. I injected an antihistamine to help purge the toxic byproducts from the animal's system. I injected painkiller, steroids to reduce inflammation in the muscles, vitamin E, and muscle relaxant, and told Duane and Lionel to feed him bran mashes and timothy hay, which are easier to digest and not as rich in carbohydrates.

"He gonna be lame now, Doc?" asked Lionel, who was giving the animal a vigorous toweling-off.

"Well, it's possible, but it's too early to tell," I admitted. "You'll have to get him on his feet as soon as possible and work him very gradually from light to moderate exercise over the next week or so.

I'll check him again in a couple of weeks if he holds his own 'til then."

I had been squatting near the horse's head and winced as I rose to my feet, pain slicing through my ribcage. Lionel jumped up to steady me. Duane, holding the lantern, shone it directly into my eyes.

"Doc, you all right?" asked Duane.

I instinctively raised my arm to shield my eyes, causing me to wince again. Duane lowered the lantern.

"I think my ribs are bruised," I said, catching my breath. "Let's get out of here before this storm gets any worse."

Lionel agreed to stay with the horse and sleep in the cabin nearby, having laid in provisions for long spells of woods work. Duane and I headed out into the worst blizzard we had seen in years.

By now daylight illuminated the knee-deep snow, but the sun was nowhere in sight. We could see only a few feet in front of us as the wind lashed icy crystals into our faces. With every step, my ribs sent a stab of pain through my chest. But I was young then and Duane was as rugged as they come. Finally, we made it out to his truck, swept off the blanket of snow with some pine boughs, and churned our way down the woods road to the main gravel road, which had received a pass from a snowplow, allowing us to forge our way to the Blue Hill village store, where I'd left my car.

Duane jumped out and helped me brush off my car. "Why don't you stop down t' the house and have a cup of coffee before you get on the road, Doc?" he asked. It sounded good. Since he lived less than a mile from the store, I accepted.

We fishtailed into Duane's unplowed driveway and stomped our feet on the porch mat, noticing a light on in the kitchen beyond the door's glass pane. "Good," Duane said, "Martha's up. Hope she's got some coffee on."

Martha did, as the aroma told us when we entered the kitchen. I waved hello to Martha, who was talking on the phone. Turning to face me, she waved back. "Oh, good," she said into the handset, "Doctor Brown just walked in." Turning to me, she covered the mouthpiece and asked, "Doc, can you speak to Lottie Leroux?"

Lottie was a client and a neighbor of the Clawsons who often brought her dogs to my small-animal hospital in Belfast. My heart sank at the thought of getting waylaid so far away from home, but I took the phone. "Hi, Lottie."

"Oh, thank goodness," Lottie said. "I just called Martha because I was so upset—I can't believe you walked in."

"What's going on?"

"Gilbert just bought an expensive new hunting dog, shipped up from Tennessee. We've only had him a few days, and now he's having fits. Can you come over? Gilbert will come and fetch you on the snowmobile, 'cause the roads are all plugged up."

"Okay," I agreed, as Martha handed me a cup of coffee. "But tell Gilbert to come quick. The road to Belfast is getting worse by the minute."

I hung up and went to stand by the woodstove, enjoying the comfort of the stove and the coffee. Minutes later, we heard the whine of the snowmobile outside.

Bouncing over the snowdrifts on the snowmobile was excruciating on my ribs, but the ride was short. Gilbert and I walked into the house. His new pup lay trembling on a blanket in the kitchen, an anxious Lottie crouching over him. Then the little dog started thrashing, his eyes closed. An examination and short discussion with Lottie and Gilbert confirmed that we were looking at some strain of distemper, one of the biggest killers of dogs. Some strains are more virulent than others. Like poliomyelitis in humans, distemper is caused by a neurotropic, or "nerve-loving" virus. When it invades the brain, the largest nerve center in the body, it causes convulsions and, usually, death. Dogs who survive any form of the disease are typically left disabled, because nerve tissue is the slowest to regenerate of all the five types of tissue that make up the body. Distemper, like polio, may wither limbs, cause muscles to atrophy, and damage organs and vital systems.

Since humans haven't yet found a cure for a virus once it invades the cells of the body, prevention is the only option. Vaccines are commonly available for canine distemper, but this puppy from Tennessee obviously hadn't received any.

After Gilbert and Lottie listened to my explanation, Gilbert asked the predictable question. "But Doc, can't you give him something to fix him up? He's a good-blooded hound and I was countin' on him to replace my old bitch, Dolly."

I explained that the only thing we could do was to medicate him to ward off secondary infections, keep him as comfortable as possible while the disease ran its course, and hope the little fellow had enough strength to survive. If he didn't improve within a week, I advised euthanizing him to spare him further suffering.

"I'm sorry," I told them, watching their faces sag. "Distemper is cruel."

Gilbert fired up the snowmobile and ran me back to my car. The storm was winding down now. By the time I brushed off my car again and headed up Route 15 toward Belfast, the plows were out, and the road was passable.

Back in my office, I gave myself an X-ray to check my ribs, and to my surprise, found that they were intact. But judging from the pain, I had torn some muscles, which would also take a long time to heal, especially when I used my upper body strenuously on farm calls every day. Luckily most of my small-animal appointments were cancelled because of the storm, and I took it easy.

About ten days later it was my unpleasant duty to euthanize the Leroux puppy. He was now having eight to ten convulsions per hour, and putting the little fellow down was the kindest thing we could do for him. After I wrote a letter attesting to the cause of death, the breeder sent Gilbert a new pup six months later.

As for Duane Clawson's horse, it survived the painful ordeal of blackwater, but his rear leg muscles never recovered their full strength, and he could no longer perform the heavy-duty woods work he was trained for. But Duane was kindhearted, and he found a part-time farmer nearby who needed a horse for light work, hauling vegetables, and he sold the weakened horse for a pittance. All considered, it wasn't a bad deal, especially for the horse.

9 vacation coverage

I had a standing arrangement with a veterinarian, Allan Jones, who practiced twenty miles from me, to cover each other's clients when either of us left town for more than a day or so. One year, during Allan's annual vacation, I got a call from a dairy farmer who was to provide me with a very memorable experience.

"Hello, Vet'nary Brown? This is Ken Ralston. Doc Jones usually does my vet'nary work, but he's away."

"Yes, I'm covering for him. What's the problem, Mr. Ralston?"

"My beef cow's been trying to calve for over a day now. The calf's front feet have been stickin' out since this morning, but she can't seem to get any further. I checked inside her, and the calf is dead. It's huge, too. A lot of my neighbors have been over here tryin' to pull her out."

"What's the status now?" I asked.

"Well, I had Charlie Buzzy over. He's kind of a homemade vet'nary. You know him?"

"Yup."

"Charlie, he spent most of the mornin' over here. He poured a couple gallons of mineral oil around that dead calf and all six of us pulled, but we didn't gain an inch. Tried block and tackle with a come-along on each foot. Still didn't gain an inch. Buzzy wanted to use the tractor to pull it, but I wouldn't let him. Finally he gave up and went home 'bout two hours ago."

"How's the cow doing?"

"She's still standin', but she's pantin' somethin' fierce. I never seen bigger feet on a calf in my thirty years of cattle farming. Looks like it was half grown already. Can you come out here?"

His place was in Orland, a river town about twenty-five miles away. "I can be there in a couple of hours," I told Ralston. "I have another emergency to take care of first, but it's in the same direction, which should save some time."

"That'll have to do, Doc. Thanks," said the farmer before hanging up.

As I was readying for the two emergency calls ahead, a physician who was a personal friend of mine came into the office and asked if I had any interesting calls on the docket. Bill Allison was single then, and he liked to accompany me sometimes on his days off.

"Sure sounds like it, Bill," I told him. "Want to come along?"

"Why not?" he answered.

We sped to the first call and found twin calves that were competing to be born first, and in the process had gotten twisted together like a Chinese puzzle. A half hour later, after grueling labor on the part of the mother and myself, we delivered a pair of healthy heifers.

Then Bill and I headed to Ken Ralston's place at a reckless speed in an attempt to gain time for the routine farm calls that I had to make after the next emergency. My old Dodge lurched and shuddered on the curves at seventy miles an hour, and I was reaching eighty-five to ninety miles an hour on the straightaways.

"God, Brad, do you drive this fast on vacation?"

"I'll let you know, should I ever go on vacation," I answered. Upon our arrival at the Ralston farm we encountered quite a reception committee—a half-dozen neighbors and hired hands, all gathered behind the patient in the barn. Among them was a "homemade vet'nary" who was stripped to the waist and had one arm inside the distressed cow.

I introduced Bill to Ken Ralston. "Pleased to meet you Doctor Allison," Ken said. "You a vet'nary, too?"

"No, I'm an M.D. in Doctor Brown's town," Bill replied.

"Oh, a mailman," Ken said. "That's a good job."

I explained that Bill was a surgeon and that "M.D." meant medical doctor.

"Well, I'll be damned," Ken mused, baffled about what would bring a physician to his barn.

"Yes," Bill said.

Behind the poor cow lay the evidence of laymen's attempts to solve a medical problem: pulleys, ropes, chains, come-alongs, soap bottles, and discarded mineral oil containers lay scattered among empty cigarette packages and even an empty whiskey bottle.

"Got troubles there?" I asked the man with his arm inside the cow.

"Hell, no," the fellow declared. "I'll have this calf delivered in ten minutes. I've never failed to get one out yet. This'll be my fifth one. Don't need you vet'naries takin' my money for this kind of work," he said with a snort of disgust.

"Go right ahead," I said. "I'm going to get my gear and some warm water in case you run into some trouble." As I departed, I said over my shoulder, "You've only got seven minutes left to deliver that calf."

Upon returning, I found that the man had given up, saying, "She's ass-wrong inside—we need a tractor."

Ken Ralston intervened. "Thanks a lot for tryin', Harold," he said. "Now let's give Doc a chance."

Harold wiped his hands on a dirty old bath towel and spit in the gutter before disappearing out the door. The other five men stood around waiting to see what would happen next.

"Ken, you were right," I said, "those feet are the largest I've ever seen on a newborn calf in my entire working life."

As the cow had been subjected to endless abuse for twenty-four hours before my arrival, I gave her a spinal anesthetic to numb the birth canal before I began my own exam. After a few minutes and a thorough internal examination, I had no doubt that the calf was exceptionally large—and dead. To try and take that calf through the normal birth procedure would mean certain death to the mother.

"Well, Ken, you have two choices for getting this calf out of her," I said. "You can either cut the calf into pieces inside or remove it through a Caesarean section. The section is her best bet, as she's so swollen and bruised from all the abuse you fellas have put her through."

Ken looked sheepish. "Well, like I told you on the phone, Doc, she's my best cow. Do whatever you think best." I nodded, wondering how he'd have treated a less favored animal.

Bill and I went to the trunk of my car to put together a surgery tray. "Brad, let me get this straight—are you really going to do a Caesarean section in that barn?" he asked.

"Yup."

"My God, I don't want to miss this. I can't believe performing abdominal surgery in a cow barn, with all the risk of infection," he said. He was almost breathless in his incredulity.

"Well, I've been doing this in cow barns for years, and so far, so good. Let's see what happens with this one."

"You do it with her standing?"

"Sure."

"I'll be damned," Bill said. "Do they ever lie down on you during the procedure?"

"Rarely."

I finished putting the instruments in the tray and did a quick inventory of everything I needed. About half an hour later, with everybody lifting, we removed a huge calf through the right side of the mother's flank. It was impossible to maintain antisepsis, as the incision was being held apart by many grimy hands with dirt and cobwebs filtering down into the abdominal cavity. I thought Bill would have a calf himself when he saw one of my helpers try to talk with a mouth full of tobacco and slop a gob of spit into the abdomen. Another fellow contributed a pack of cigarettes from his shirt pocket when he bent to help. I retrieved it.

After extracting the enormous calf I started suturing the uterus. "Ken, can you go to the house and get a box of salt?"

He soon returned with it, and I prepared a gallon of saline solution. I added a couple grams of the antibiotic tetracycline and dumped the mixture into the peritoneal cavity and sewed up the abdomen. Then I donned a plastic sleeve, removed the placenta vaginally, and packed the uterus with antibiotic boluses.

At this point Bill was flabbergasted. Ken went off to get a pair of scales, saying that he wanted to weigh the calf, the largest he'd ever seen. "Wow!" Ken exclaimed, when the dead animal was on the scales. "Two hundred and twelve pounds. No wonder the mother couldn't have it. Weldon, go over to the house and get the camera." He turned to me and thanked me for the "fancy operation." I told him I was glad to help and instructed him on how to care for the cow for the next few weeks.

Bill and I started cleaning up, but I noticed that Ken was still standing beside us, shuffling his feet. Finally I looked up at him.

"Uh, while you're here, Doc," he said, "I got three or four cows I need checked for pregnancy. And I've got a couple hard milkin' teats I'd like to have you open for me. Think you can do that?"

"Okay, if we make it quick. I've got a lot more calls to make."

The second cow to be checked for pregnancy was stanchioned with her hind end facing a steel post, which was not uncommon in the type of barns that had been built in the last twenty years. With my arm inserted up to the armpit in the cow's rectum, I gambled that the cow would stay on one side of the post long enough for me to get the job done. In spite of the cow's being tethered by the neck, a person in my position could easily fracture or dislocate an arm if the animal twitched her hindquarters. Over the years I had done thousands of rectal pregnancy checks on cows and horses without serious injury, though I could recall many near-disasters. This particular cow was nervous, and my luck ran out as she slammed her hind end against the post, bending my elbow in the wrong direction. The pain was momentarily unbearable until she careened into the opposite direction, freeing me and leaving the joint partially dislocated. I stepped back and withdrew my arm in a hurry.

Bill had gone to another barn to look at the young cattle, but Ken was standing beside me. I told him to grab my hand and jerk it quickly. A searing pain was rewarded by the pop of the joint snapping back into place. I perspired and swore like a pirate, but Ken seemed unphased. The incident left me feeling a little skittish as I continued working.

We finished the pregnancy checks and moved on to the cows with the hardened teats. As often happens in dairy cows, these teats had been clogged by scar tissue that prevented milk from flowing through. The scar tissue forms when the teats are irritated, and milking machines do a good job of irritating them. Not only does a blocked teat make milking painful for the cow, but it also traps milk in the udder, putting her at risk of mastitis, or infection of the mammary tissue. To enlarge a teat orifice, I used a small tool especially designed for the task, which caused little pain for the cow because scar tissue contains few nerves.

I had finished one teat and started on another when my patient hauled off and kicked me in the forehead. I can't say as I blamed her, and luckily it was a glancing blow, leaving only a lump and a small cut. Blood was running into my eye when Bill returned, just as we were finishing up.

"Hey, Doc," Ken said to Bill, "we could've used you a few minutes ago. Doc Brown here nearly broke his arm before he got that gash in his head."

Bill shook his head as he swabbed the cut with some rubbing alcohol from my grip. "Well, Brad, it looks like you're going to live." We gathered up my gear and took off for the next call.

The day continued with a string of emergencies that my office passed on via my car radio. Bill had seen all manner of conditions and treatments, but the Caesarean section in the barn really stuck in his mind. We were returning home around 2 A.M. when he said, "I'm willing to bet that cow won't live for a week, and if through some miracle she does, she'll never conceive and give birth again. She's bound to be rendered sterile from chronic endometriosis," he pronounced, referring to an infection of the uterine lining.

"What'll you bet, Bill?"

"A half-gallon of Pinch scotch if she lives one week and an added gallon if she calves again," he said.

"You're on."

When Allan Jones returned from vacation, I learned why Ken's calf was so exceptionally large. On the day of the Caesarean I had repeatedly quizzed Ken if the cow was overdue. He kept insisting she was at her usual nine months, but he admitted he'd lost the paper documenting when she was bred. Allan called the artificial insemination technician, who found his carbon copy and confirmed that she had been nearly three months overdue at the time of the Caesarean section. The prolonged pregnancy was caused by a rare genetic abnormality known as fetal gigantism.

Normally each sex contributes a gene that regulates the length of gestation, or how long a fetus develops inside the uterus. In the overwhelming majority of pregnancies, both genes specify the same gestation period. But in rare cases one gene calls for a longer term and dominates the normal gene. Instead of being born on schedule, the fetus continues to grow until it outstrips the mother's capacity to supply nutrients through the umbilical artery. Then the fetus dies and the mother can't deliver the enormous stillborn.

When Allan told me about the insemination date, I asked him to keep me posted on the cow's health and whether she was able to deliver again.

Fourteen months later, Allan gave me a call: Ken's favorite cow had delivered a hundred-pound heifer calf—without any surgery. The drinks were on good Doctor Bill.

During the drive home after our little celebration, I realized that a vague discomfort had haunted my evening. It was an increasingly familiar reflection about the domination of animals during the age of mass-production farming.

When I was growing up on a farm in the 1930s, the average life of a dairy cow was about eleven years, during which she produced about six thousand pounds of milk a year. She had a name and was a beloved member of the family. Today's cow lives an average of seven years, pumps out at least eighteen thousand pounds of milk annually, has a number instead of a name, and is treated like a machine. Her body is the result of countless generations of breeding aimed at producing more and more milk. An exceptional dairy cow today can put out up to fifty thousand pounds a year—twenty-five tons. That breaks down to a little less than a hundred and forty pounds, or seventeen gallons, a day.

At the end of my practicing years—almost three decades ago—farmers got about $1.15 for a gallon of raw milk from a middleman who pasteurized, bottled, and distributed it. The milk buyers also processed various other dairy products, like skim milk, cream, and cottage cheese, in the end making about five times per gallon more than they paid the farmer. After expenses and depreciating equipment, farmers barely made enough money to survive. Such disparities have doubtlessly widened today, shedding light on why so few small dairy farmers have remained in business.

Our society often prides itself on developing technologies that deliver more food, faster. But more and more often, our "progress" comes at the expense of the animals we rely on for food. For one thing, high milk production puts stress on the cow's body, making her vulnerable to secondary infections. And high-producing cows are hormonally imbalanced, which makes it hard for them to get pregnant. This has evolved into another entire business for farmers, vets, and breeders. Because such cows are worth thousands of dollars or more, vets administer hormones to get the cow to superovulate, or produce as many egg cells as possible. Then the vet removes the eggs and surgically implants them in surrogate mothers. At every step, someone, including the vet, is making money.

Farm Calls

Photographs by Lou Garbus

Disinfecting before delivering the next calf

A breech birth. The calf should arrive nose and front feet first.

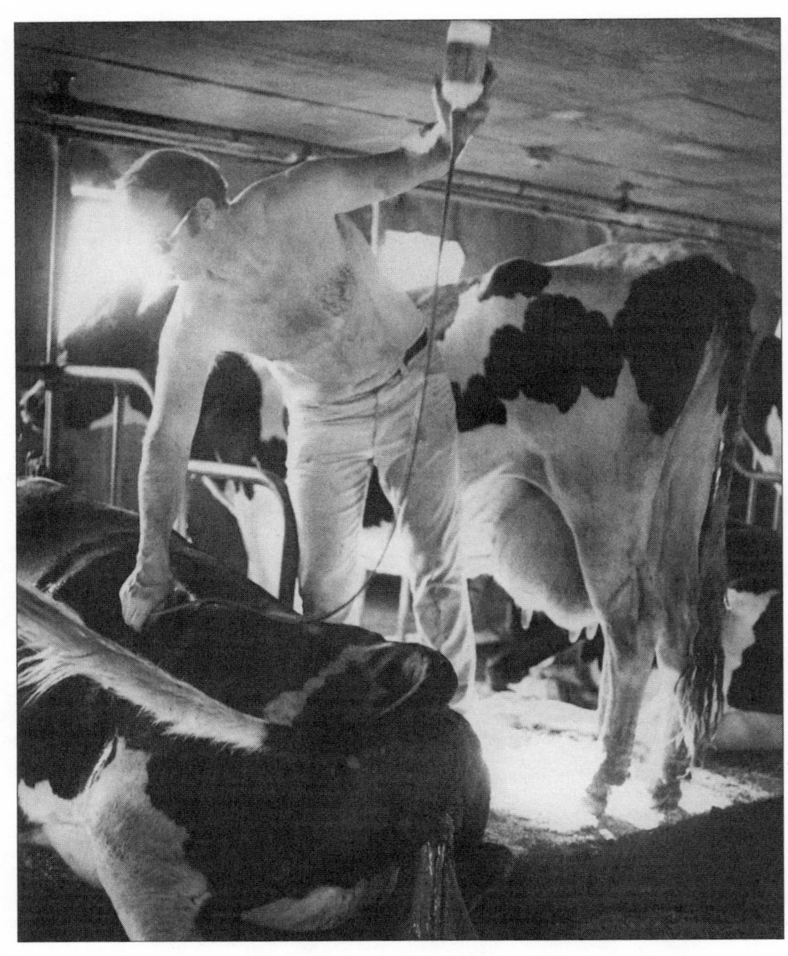

A hypo-calcium treatment for a cow with milk fever at Littlefield Farm

Kids meet the new calf at the Lee Larabee farm in Knox

Repairing an umbilical hernia

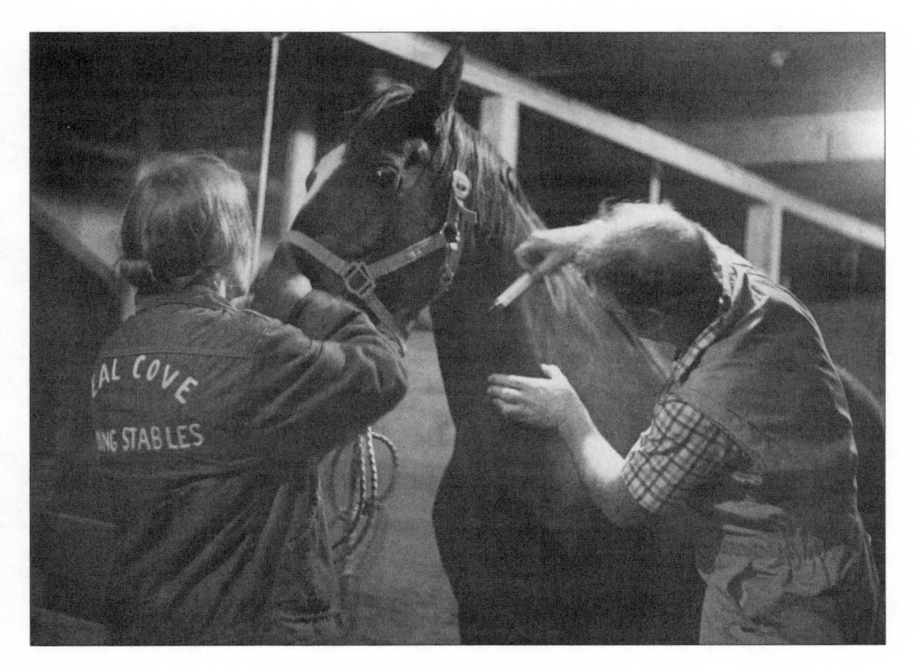

Testing for Coggins at Seal Cove Stables

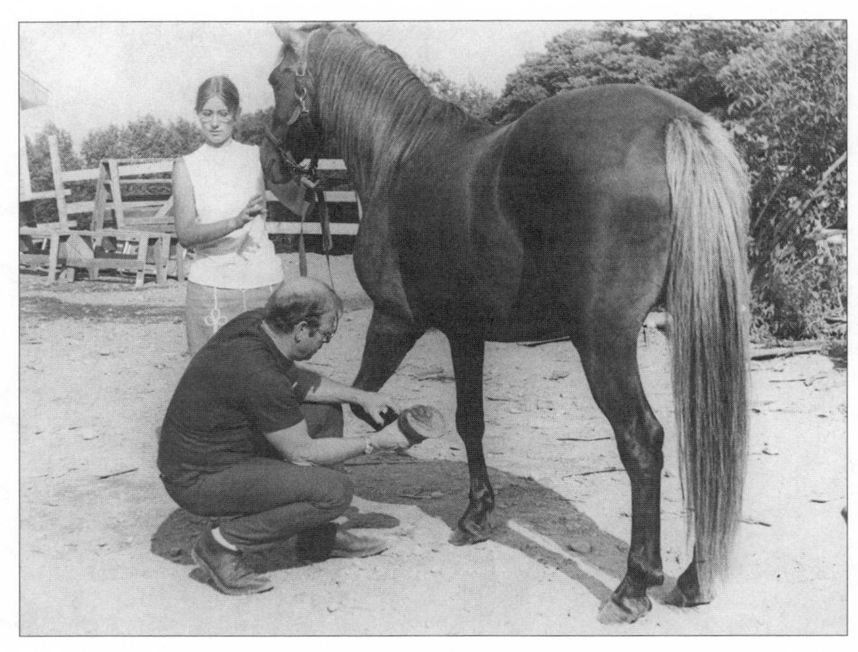

Checking Millie for lameness at Moffit Farm in Liberty

A quiet moment, waiting in the tie-up for the next horse

Preparing for surgery

A sick calf shows up for an office visit

10

A Day at the Joy Farm

It was six o'clock in the morning. On my examining room table lay a handsome two-year-old collie named Rex, who had been shot in the head with a 32-caliber rifle an hour earlier while paying a visit to a neighboring female in heat. The bullet had entered the left nostril and traversed the calvarium—the domelike portion of the skull—before shattering into several pieces. The fragments were now lodged at the base of the brain's frontal lobe. Luckily, they had missed many vital centers that control memory and other cognitive functions. With the aid of X-rays, I determined that I could remove the bullet fragments and perhaps restore Rex to a normal life. The surgery was risky, but the other option was death by euthanasia. Rex's owner had agreed that we should give it a try.

About an hour and a half later I had succeeded in removing the life-threatening pieces of lead from the brain floor and injected Rex with antibiotics and tetanus antitoxin. I was just stitching up his head when the phone rang.

"Doc, this is Mabel Joy," said a stressed female voice. "Please come out here as quick as you can. We've got a cow down with milk fever and she's cast her wethers. She's our best milker."

"Okay, Mabel," I said. "I'll be there in half an hour." I hung up, carefully laid Rex in a large kennel where he could safely emerge from anesthesia, and left a note for my assistant. Then I loaded my car with the supplies I'd need and headed out full speed ahead to the Joy place in Waldo.

Milk fever is an emergency with a capital E. The scientific name for the condition is parturient paresis. It isn't actually a fever at all, but a chemical imbalance that triggers a potentially lethal chain of events. It can occur in cows of any age but is most common in mature, high-producing dairy cows, five years old or older. A cow's bloodstream maintains a delicate balance of calcium and phosphorus. Immediately after calving, she develops a large udder to feed her calf, and the udder sometimes stores too much milk. When this happens the first milking or nursing can drain too much calcium from her bloodstream. Her muscles weaken and she collapses. Paralysis and coma can follow in short order, usually leading to death within hours after the first symptoms appear.

I arrived at the Joy place within thirty-five minutes. Mabel and her husband, Ezra, met me in the barn, where their prize milker sprawled in her stall. She was already comatose, rolled partly on her back with her head on one side of the manure gutter. I immediately injected an intravenous solution of calcium, phosphorous, magnesium, and dextrose, inverting a 500-cc bottle and draining it into her jugular vein through a rubber tube running between the bottle and the needle in her neck. Mabel held the bottle while I turned my attentions to the ejected uterus, which lay on the floor behind the animal.

"It looks pretty bad, Doc," Ezra said.

"Sure does," I said, shaking my head.

Milk fever is often compounded by an everted and prolapsed uterus, known in farm vernacular as "cast wethers," also caused by the drop in calcium. One of the things calcium does is help maintain muscle tone. When levels of this vital mineral plummet, the uterus, an organ made entirely of smooth muscle, suddenly weakens. The strain of giving birth can turn it inside out like a sock (eversion) and cast it out through the vagina (prolapse). This is no small problem: normally weighing only twelve to sixteen ounces, the uterus swells to forty to seventy pounds during pregnancy. The enormous organ dangling outside of her body can throw the cow into shock and damage vital arteries. Her stablemates may also trample it while the stricken cow is laid out on the floor.

In this case the cow's uterus was still tethered to her vagina by the mesometrial ligaments. Because she lay with her rear quarters slightly downhill, I quickly saw that it would be impossible to squeeze the swollen organ into her from this angle. I needed gravity on my side.

"Come on, Ezra, let's get her all on one level," I said.

Grunting and puffing, Ezra and I dragged the unconscious cow back onto the level barn floor. I kept hoping that she would revive enough to stand, which would make my job a bit easier. After receiving the intravenous fluids, a cow can revive in a seemingly miraculous thirty minutes. But this old girl was a slow responder.

Squeezing a uterus back inside a cow is like stuffing an orange through a keyhole. It usually requires stripping to the waist and lying prone on the barn floor in a fetid pool of afterbirth fluid and blood, not to mention manure. This case was no exception: I peeled down the top of my coveralls and lay down in the muck. I rinsed the uterus in a disinfectant solution to remove any remaining pieces of placenta, which could cause infection. The organ was cut in several places, no doubt having been stepped on by her stablemates, so I sutured it before giving her a shot of pituitrin in the leg muscle to shrink the smooth muscle of the uterus. In the meantime, I began to massage it gently, pressing fluid out of the tissue to help reduce the swelling. Once these measures had taken effect, I was able to turn the battle-scarred organ right side out and work it back into place. Then I gave her more injections to slow hemorrhage, speed up her recovery from shock, and facilitate smooth-muscle contraction.

"You think she's gonna make it, Doc?" Mable asked.

"At this point, I don't know," I answered. "We've done everything humanly possible for her, and now we're just going to have to wait."

I went to wash up in the milk room, with Ezra shuffling behind me. He watched me for a few seconds before clearing his throat. "Er, while you're here, Doc," he began, "I've got six Herefords I just bought from a neighbor. They're s'posed to be pregnant. He had them with the bull for a couple of months. Thought you ought to check to see if that bull got 'em with calf, 'cause he was young and unproven."

"I guess we can take a look, as long as they're rounded up," I replied. I dropped the scrub brush back into my pail.

"They're down to Father's pasture," Ezra said. "I've sent the hired man down with a bucket of grain to get them into the corral."

I had a waiting room full of people at the animal hospital, but I knew that this task, as with most "while you're here, Doc" jobs, would take longer if I postponed it until the next time I happened to be in the area and no one was around to help. And now here stood Ezra, a willing volunteer.

"It won't take too long," he promised as we headed out the door. Nearly every time I heard those words from a farmer, I got a sinking feeling in my stomach.

Ezra's truck bumped down the dirt farm road ahead of me for about three miles. Then I spotted the crude corral Ezra had hastily assembled in the corner of the pasture next to the road. The structure would have sent a western beef rancher into fits of laughter. It was constructed with snow fencing the state highway crews had discarded the year before. In Maine and other states that get heavy snow, the highway department puts up lightweight fencing in winter to keep snow from drifting into the road. Nowadays it's nylon webbing, but then it was made of rough wooden slats wired together. Ezra had built his corral eight feet high and about a hundred feet square. The "gate" consisted of a pickup truck the hired man had parked across the opening.

The Hereford is a typical beef cow—it's perfectly docile until you put a rope on it or disturb its environment. Unlike dairy cows, most beef critters have never felt a restraint of any kind, and they don't enjoy the experience. Agitated beef cows intend to remain free, and, if given no choice, would just as soon trample anyone or anything that gets between them and freedom.

What happened next proved me right. I entered the flimsy corral, carrying a box of plastic rectal sleeves and a thirty-foot nylon rope with a metal snap-release on one end. The suspicious cows swirled about the small space, each weighing at least a thousand pounds and equipped with a pair of sharp, two-foot-long horns. My job was to lasso each one and secure her to the bumper of the pickup truck. This was the rodeo part of the pregnancy exam: once the thirty-foot rope tightens around the neck of one of those critters, she becomes a half-ton fish out of water, leaping eight to ten feet in the air in all directions. To hang on to the other end of the rope you have to be as agile as a goat. It also helps to be young.

Ezra's hired man held each cow steady while I plunged in, so to speak, bracing myself against the cow's hind end and keeping an eye out for her hooves while I tried to conduct my delicate task. The actual pregnancy exam took only a few minutes. I stood behind the cow and inserted my plastic-sleeved arm into the rectum to the armpit, and very carefully felt for the uterus, which, in the early stages of pregnancy, lies beneath the rectum in the pelvic cavity. At less than thirty-four days, the embryo was a BB-sized bump in the endometrium, the mucous membrane lining the uterus. After my fingers detected that, I gently manipulated the ovaries to verify the presence of a large corpus luteum, or "yellow body," another indicator of pregnancy. I could usually determine pregnancy down to twenty-eight days. Embryos older than thirty-four days were easier to detect, as I could "slip" their fetal membranes between thumb and forefinger like the skin of an empty balloon. In this herd the longest-term pregnancy was around sixty days and the shortest about twenty-five days.

I thoroughly enjoyed the OB-GYN discipline of veterinary medicine, because it combines hands-on skill coupled with reams of textbook knowledge. Had I wanted to move to an area with large herds of cattle or horses, I would have made it the sole focus of my practice.

We lassoed and checked five of the six subjects and found them all to be pregnant. Past experience had taught me that the last cow to be caught is the most distraught and will fight the hardest. But this time, despite her jumping around like a parched pea in a hot skillet, I managed to lasso the last cow without much trouble. As I'd done with all the others, I ran for the truck bumper with my end of the rope. Then, somehow, before I got there the cow managed to tangle the rope around my right thigh. Two seconds later she crashed through the flimsy fence of the corral, sending parts of it twenty feet into the air.

The hysterical beef critter galloped across the pasture, dragging me along the ground. Bouncing like a ball on a chain, I fought to stay face-up, partly to try to reach a sitting position that would allow me to haul up enough slack to free my leg, and partly to avoid shredding my face on the rough ground. But her constant flight kept the rope as taut as a cable, giving me no chance to reach even a half-sitting position. I didn't dare to dig my other leg into the ground, knowing it would fracture immediately, as we were doing at least fifteen miles an hour. So I tried to flex my free leg to dig my heel in, knowing I'd have to stay on my back in order to survive.

These thoughts raced through my brain as my body was pummeled by rocks, discarded fence posts, and hummocks of hard earth. It took all my strength and balance to stay on my back, and more than once I got flipped onto my stomach, my face bumping through rocks and dirt. I could feel rocks chipping my teeth and clots of dirt filling my mouth between breaths of precious air. Pain tore through my thigh, and I could no longer feel the lower part of my leg. I was getting pounded into a state of semi-consciousness, with no way to see that Ezra and the hired man were running after us.

What I had realized from the moment of take-off was the heifer's direction—straight for the woods. Even if I survived the two hundred yards of open pasture, I had no doubt that once she reached the tree line, my life would be over in a matter of seconds—and it wouldn't be painless. As my body scudded toward the looming edge of the woods, one thought kept cycling through my mind: "This is a hell of a way to go."

About thirty feet from the tree line, a miracle happened. The cow slowed her thundering pace, which gave me my one chance to sit up, grab the taut rope, and dig in my heels. It worked. Even though she caught her breath and surged ahead again, the momentary pause gave Ezra and the hired man time to catch up. They pounced, grabbed the rope, and I freed myself in a flash.

Relief flooded in, and so did the pain. I felt like I had just crawled out from underneath a steamroller. Blood covered my face and head, my nose was broken, and my right leg was swollen twice the size of the left. I knew at a glance that the constriction and twisting of the rope had ruptured several leg muscles above my knee.

Holding fast to the panting cow, Ezra huffed for a few seconds before wheezing, "Gawd, Doc, your face looks like it's been put through a meat grinder."

"Thanks, Ezra," I wheezed back. My coveralls were in tatters, but I was alive. When I was able to stand, I decided this ordeal wasn't going to be for naught. I limped to the exhausted beef critter and checked her for pregnancy. She proved to be the only one that was "open"—that is, not pregnant.

"It was good we caught up with ya, 'cause I'd a thought she was pregnant, seein's how you found the rest of them that way," Ezra drawled. He allowed that catching up with me had helped me out as well.

"Got that right, Ezra," I said.

"Uh, how much I owe ya for the pregnancy checks, Doc?"

"Dollar a head, Ezra."

"Well, I reckon you earned it today."

I had no response for that. We hobbled back to the dooryard, where I slid my aching body into the car and headed for my office. I washed the open wounds on my arms and face and stitched some of the worst cuts. I changed my clothes, gulped a cup of coffee, and wrapped my damaged right thigh in an elastic bandage. Where the muscles were torn, my skin had turned an ugly purple, and the rest of my legs and back were blotched with yellow and brown. The worst pain, I knew, had yet to set in.

I checked in on Rex and the other recent surgery patients. All were doing well, and Rex was awake and wagging his tail when I stopped by to say hello. I went back to my office and sat there awhile with my leg propped on the desk, feeling the swollen thigh begin to throb. But I was alive, and if the pain was a reminder, so be it. Luck had been with me again.

11

Chester

One April morning Phyllis Harper called from Lincolnville, a coastal village about fifteen miles from my vet hospital in Belfast. "Hello, Doctor Brown," she said. "Listen, when you're out this way, could you stop in and take a look at Chester, our new pig?"

"What's up, Phyllis?" I asked.

"Well, we bought him at auction when he was a piglet, thinking we'd raise him for eatin'. Only thing is, he was s'posed to be castrated at four weeks old, but he's far from that now and his bag's the size of a softball," she said. By "bag," she meant Chester's scrotum.

"So now he needs castrating, so that his meat will be sweet?" I said, wondering why she was taking so long to get to the point.

"Well, yeah, except Cy"—her husband—"asked our neighbor, Frank Auld, to come and look at Chester's bag, seein's how it looked kinda big for such a young fella. Me and Cy don't know nothin' about pigs."

Anything but call a vet, I thought to myself. "And what did Frank say?" I asked, as patiently as possible, given that I was trying to prepare for my annual tax meeting with my accountant, on top of the usual list of farm calls and small-animal surgeries.

"Frank reckons Chester was born busted, which would explain why we got such a deal on 'im—damn auctions," she said in a tone of disgust. "Anyway, Frank says if he's busted, he might still have one nut left up inside 'im, so even if he's castrated we can't eat 'im. That true, Doc?"

The condition she described was an inguinal hernia. In general, a hernia is a condition in which part of any internal organ protrudes through the wall of the cavity that contains it. An inguinal hernia is an opening in the peritoneum, the thin membrane that lines the pelvic and abdominal cavities, containing the intestines and other vital organs. The opening, caused by a rupture or a minor birth defect, allows abdominal fluid or even part of the intestine to drop into the scrotum, causing swelling and pain, and sometimes pinching off the intestine, which can be life-threatening. If the hernia results from a birth defect, sometimes one of the testicles remains in the pelvic cavity, where it developed during the mother's pregnancy. During the normal development of a male fetus, both testicles descend from the abdominal cavity into the scrotum before birth. If the undescended testicle, called a cryptorchid, remains for a couple of years it becomes what is known as a sertoli-cell tumor, a benign body that secretes the female hormone estrogen, causing the animal to become feminized. Usually a hernia, the removal of the cryptorchid, or both can be repaired in a simple operation.

"It's possible, Phyllis, but I should take a look at Chester before we can really say. I have to drive by your place this afternoon, so I'll stop in," I said.

After my small-animal office hours and surgeries, I headed out with a list of five "when by" farm calls in the area near my accountant, with Chester the pig at the top of the list. The Harpers lived about fifteen miles from my office, so, while driving, I had time to spread the reams of tax documents my bookkeeper had given me all over the front seat and floor of the car, trying desperately to find the ones she had instructed me to look over before meeting with the accountant.

Five miles into the drive it started raining tumultuously, and it kept up as I drove into the Harpers' yard. It wasn't a real farm, just a house on three or four acres with a few small outbuildings where Phyllis and Cy, a postmaster, raised a few chickens—and more recently, pigs. Before I reached the kitchen door to announce my arrival, my clothing was soaked through to my hide. After several knocks, Phyllis came to the door, invited me in, and beckoned me to the kitchen window. She pointed to a small building in the field.

"He's probably inside his little house out there, Doc. He could be roamin' around somewhere, but I doubt it, where it's raining so hard and all."

"Naw," I kidded her. "He's probably out sunning himself."

"Get outta here, Doc," she said, chuckling.

I set out alone through the field, which the rain had turned into a swamp. It was now growing colder, and the chill seemed to sink into my bones by the time I reached the pig house. The crude shelter sat in a fifteen-by-fifteen-foot yard enclosed by a makeshift fence of orange and grapefruit crates. I pushed aside one of the crates, entered the pig's domain, and pulled it back into place behind me. Then I opened the door of the shack to get my first look at Chester the pig.

The animal, which appeared to weigh between two hundred and three hundred pounds, looked up from his bed of hay to stare at me with intelligent, beady eyes. He pointed his moist snout in my direction and sniffed, as if trying to figure out whether I was friend or foe.

"Hello, Chester," I said in my most reassuring tone. I had acquired the habit of talking to animals, in the hope that the vocal tones, not the words themselves, would soothe them. I had learned that very few animals, especially large ones, were happy to see me, especially when I trespassed onto their turf. As I stooped to approach him in the low-ceilinged shack, Chester scooted to the opposite corner, and I caught a good look at his "bag," as Phyllis called it. Sure enough, Frank Auld had been right about the hernia. Chester was like so many auction animals sold at a bargain with a hidden price attached. Surgery would be required to repair the hernia, and while I was inside I could remove his testicles.

Since the problem required little examination to know what had to be done, I shut the door and ran back to the house through the drenching rain. Phyllis was making bread in her warm kitchen. I explained that I needed to remove the runaway testicle and repair the hernia, as her neighbor had suspected.

"How much?" she asked.

"Twenty dollars," I replied.

"OK," she said. "When can you do it?"

"Right now, if you can give me a hand."

She hesitated. "I've never done anything like that," she said.

I explained that she would need to hold a rope on Chester until the anesthesia took effect. After that she would hold another rope on his leg to keep it spread apart so I could operate. She could turn her head and look the other way if she wanted, I reassured her. Nodding, she put on her slicker and walked with me to my car, where I got the necessary equipment out of my trunk and dropped it into my stainless-steel pail: disinfectant, scalpel, Sodium Pentothal, clean towel, suture, needle, syringes, and vials of injectable antibiotic.

When we reached Chester's house, he had settled back into his bed of straw and fallen asleep. I was heartened, because naturally it would be easier to get a jump on him if he was in dreamland.

"Quick—shut the door," I whispered to Phyllis, who crouched next to me.

Chester's beady eyes snapped open. With a grunt, he shot to his feet and beat Phyllis to the door, charging into the rain.

"No problem," I said. "I'll drive him back in. Stay here."

But after five minutes of chasing Chester around his fenced yard, I got that familiar sinking feeling I knew from long experience in dealing with pigs, the genius of the animal kingdom. Chester had smelled the medicines strange to his environment, noticed two humans creeping up on him, and figured that more room would give him the security he needed at this time in his life. I called to Phyllis, who was staring out at me from the door of Chester's house, that she should come out and bring the pail. We'd be doing the job outside if necessary, rain or no rain, as time was fleeing.

How to describe an outdoor pigpen after it has been raining for a couple of hours? Given half a chance, pigs are the cleanest of all animals. But confined in a small outdoor pen that had never been cleaned, Chester hadn't been given that opportunity. The rain had mixed eight inches of soil, decaying table scraps, and feces into a fetid quagmire the consistency of pudding.

Squinting through the downpour, I looked around to see where to tie a rope, once I'd secured it to one or both hind legs. The flimsy crate fence wouldn't hold, so I chose the upright strut of a paneless window frame in Chester's house.

Though I'd castrated much larger pigs than Chester, he was the most nimble one ever to challenge my catching technique. Rope in hand, I wildly pursued him through the stinking quicksand, but I seldom saw more than his back, his stout little legs churning the quagmire. He lurched and grunted his way through the muck as I submarined my hand and arm in the direction of his hind legs. But not once was he tempted to seek refuge inside his home.

"Come on outta there, Doc," Phyllis shouted through the deluge. "You're gonna drown!"

I didn't care, because it seemed that at last Chester was beginning to tire. I waded in and grabbed him. Gripping one of his hind legs, I looped the rope around it with my free hand. But my plan was not to be. The second he felt the rope tighten around his hock, Chester received a jolt of adrenaline. For the next five minutes I became a pig-yard water skier—on my belly. Once in awhile Chester would pause long enough for me to lift my head above the unspeakable mire. From what seemed like a great distance away I could hear Phyllis shouting, "Let go, Doc! You're gonna drown!"

Finally, I lost my grip on the rope, and Chester kicked free of it soon after. My vision marred by mud and rain, I spotted it floating near the fence, where Phyllis could easily reach it. Once again, Chester was slowing down, huffing clouds of hog breath into the cold air.

"Quick, Phyllis—get me that rope!" I shouted, struggling to my feet.

She tossed it to me. This time, when I reached Chester's back end, he put up only a half-hearted struggle. I managed to secure the rope to his leg and tugged the exhausted pig toward the door. He floated through the soup like an enormous sausage, and I shoved him inside his living quarters. Phyllis was right behind, and this time she didn't need to be told to shut the door.

Raising the pig's posterior and tying the rope to the window frame, I quickly washed and disinfected my hands and gave Chester a shot of sodium Pentothal in an ear vein. He dropped into sleep. I asked Phyllis to hold a flashlight I always kept in my coveralls, and it was quick work to remove his testicles—the one in the scrotum and the one tucked up near his intestines—and repair the hernia.

We sat there as I washed up, the rain hammering the roof of the pig house. About half an hour later Chester awoke, shook his head, and staggered over to his bed to sleep off his drunken state.

As we walked back to my car, I instructed Phyllis not to slaughter him for at least two months, as the male hormones from Chester's testicles wouldn't leave his system until then.

"You wouldn't want to defeat the purpose of my swimming lesson, would you?" I kidded her.

"God, Doc, I thought as like you'd drown," she said, as I brushed mud from my coveralls. The rain, at last, was fading to a drizzle. We reached my car and I lifted my pail to put it back in the trunk.

"Hey, Doc," she added. "While you're here, you wouldn't take a look at my chickens, would ya?"

As we walked around the back of the house, she explained that her hens had stopped laying eggs. One look at the windowless, ten-foot-square chicken house told me the problem.

"Phyllis," I said, "you need to put a light bulb in here. In all animals, even people, reproductive hormones are stimulated by light. It comes in through the eye to the pituitary gland in the brain, where the hormones are located and activated by light. Eggs are a result of that process."

Phyllis nodded, fascinated. "God, Doc, I had no idea."

"Why don't you put in a hundred-watt bulb and leave it on from sunset to sunrise every day and see what happens?" I suggested. "And have Cy put a couple of windows in here to take advantage of the natural daylight."

"OK," she said. "I think he was afraid of foxes and coyotes gettin' in."

"Well, foxes can smell them anyway, and if you put in glass they can't get in."

As I headed down the road to my next call with the heater blasting, my car began to reek of Chester's yard. I could feel the stench emanating from my birthday suit and two layers of clothing. Even after I switched off the heater and rolled down the windows, the odor was unbearable. I stopped at the first brook I saw beside the road and stripped to the waist, washing off the solid matter. By the time I climbed back into the car, my body was shaking with the first stage of hypothermia. But with the help of the heater, the trembling finally abated and I was able to finish my other calls uneventfully.

At last I headed for my accountant's office. I noticed that Gladys didn't give me her usual peck on the cheek, and she looked none too pleased when I handed her a sheaf of puckered documents spotted with feces and mud. She conducted our business at a record pace.

Months later, Phyllis Harper stopped by my vet hospital with two dozen eggs and a slab of bacon from Chester. The bacon and eggs idea was witty, but I ate only the eggs and gave the meat away. Even after our tussle in the mud, I couldn't bear to take a bite out of the animal I had helped to repair.

12

A Cold Night on Muzzy Ridge

One bitter cold evening I was called to examine a draft horse with colic, a condition well known to horse owners. The owner, Hersey Clements, had sounded upset on the phone, telling me the horse was in terrible distress and to "come quick."

Because of their enormous large intestines, horses experience more than their share of digestive disturbances, from enteritis (diarrhea) to increased peristalsis, or the too-rapid squeezing of food through the bowel. Their long intestines also are prone to becoming twisted, blocking the passage of waste—a potentially lethal condition. Any of these malfunctions can produce symptoms grouped under the heading of colic: cramping and pain, sometimes signaled by fidgeting rear legs. Its severity depends on where the malfunction lies and its underlying cause. So the label didn't give me much of a clue about what might be at the root of the horse's discomfort.

The country roads were ice-covered, slowing me down considerably during the thirty-mile trip to Muzzy Ridge, but as this was a fire-engine emergency, I drove as fast as I dared the whole way. The last two miles were a nightmare. Hersey lived on a steeply pitched, mile-long side road coated with glare ice. Going in the journey was all downhill. With all the courage I could muster, I started down the road, but within a few hundred yards the car went into an uncontrollable spin toward the ditch. I gunned the engine and burst through the snow bank lining the road into the bordering field, keeping the accelerator to the floor. Fortunately, six inches of hard-packed snow on the field gave me some traction. So I headed down the field, driving as fast as I could, spewing a plume of snow in my wake. This was in the days of rear-wheel drive, and although my car had studded snow tires, if I had let up on the accelerator for a second, I'd have been stuck.

My headlights picked out a lonely little house ahead. Hoping it had a plowed driveway that would allow me to get back to the road, I barreled straight into the yard. After taking down the clothesline, I shuddered into a wide turn and sailed back onto the road. I skidded over an icy wooden bridge with the smell of burning rubber in my nostrils. With my gas needle showing only half a tank, I shimmied and fishtailed—uphill now—to Hersey Clements's little farm.

Hersey had sanded his dooryard, at least, so I came to a stop without hitting anything. A frail man in his seventies, he came out of the house to greet me. We went directly into the tumble-down shed where he kept his "woods horses," which is what most farmers called their draft horses because they used them to haul logs out of the woods. By kerosene lamplight, I got the first look at my patient.

"Dick ain't eaten a blasted thing since yesterday mornin'," Hersey said. "He's been gettin' up and down all day and rollin' somethin' fierce when he's down. I came up from callin' ya and he's been standin' there quiet ever since."

"Have you given him any medicine, Hersey?" I shined a flashlight on the horse's head and saw fluid streaming from his nostrils. I put my hand on his carotid to take his pulse and placed a stethoscope on the side of his chest.

"Ned Johnson come by and give him a shot in the neck of that new medicine—penicillin, or somethin'," Hersey replied. Johnson was one of many amateur veterinarians who sometimes preceded me on emergencies. "Said he'd be fine by tomorrow. Guess it's workin' faster than he thought."

I looked Hersey right in the eye. "I have some bad news," I told him. "His stomach is ruptured, and he's going to die very shortly. You had better let me put him to sleep, unless you want to see him die a violent and painful death."

For reasons I couldn't determine without surgery, I knew the animal's stomach had burst when I saw the fluid oozing from his nostrils. It was the contents of his ruptured stomach. Because of their long esophagus, which lacks the strength to propel food from the stomach back to the mouth, horses can't vomit. So when you see stomach contents flowing out of the nostrils, you know it's all over.

Hersey looked stunned. "Go on, Doc," he said. "He's a lot better now than he's been all day. He's stopped his thrashin' and rollin'. He'll be as good's new in the mornin', like Ned said."

"Well, I suggest you take him out of this shed right now, or he'll stave it to smithereens."

Hersey set his jaw. "Naw. He's all right right where he be."

"Hersey, that horse is your property. But you shouldn't have the right to subject him to what you're about to witness."

"I ain't payin' you or any vet'nary to come down here and tell me to put my horse to sleep."

No sooner had the words left his mouth than the horse uttered a hideous gurgling sound, reared up, and fell over backwards, crashing through the back wall of the shed. After crashing amongst the debris for a minute or so, he left his pain and suffering behind.

We stood silent for a moment. Hersey sucked his teeth. "Should a believed ya, Doc," he said at last. "Next time I'll put more stock in your judgment."

Disgusted, I gathered up my gear. Hersey cleared his throat. "While you're here, Doc, maybe you could check my other horse's teeth. He's been chewin' funny for six months or so, spittin' out his food now and then."

"Where is he?"

"I put him in the other barn when Dick got to rollin' and thrashin' today."

We walked without talking to the next building, snow squeaking under the rubber treads of our boots. Inside the small barn the air was only slightly warmer than outside, but the plank walls at least stopped most of the wind. Getting out my flashlight, I pulled the horse's tongue aside and spotted several elongated molars on both sides of his mouth. I gave him a sedative, rummaged in my grip, and inserted a speculum to keep his mouth open. Then I rummaged some more in search of my dental cutters, explaining to Hersey what I was about to do. I told him no painkiller would be needed, because horses have a very thick layer of enamel, which, unlike the underlying layers of dentin and pulp, contained no nerve endings. I would be trimming only the enamel.

"Huh," he said. "What made them back teeth longer?"

"Well, he has an overbite, which causes him to wear down the inside surface of his back teeth more than the outside. Over time the outside becomes too high as the inside surface wears down."

"Did you ever have to pull a horse's tooth, Doc?"

"Sure," I replied. "Last month I pulled an abscessed tooth out of John Miller's old work horse. He had a swelling under his left eye bigger than a softball—very painful. I had to drill through the bone into the sinus and pound the tooth down to loosen it before pulling it. In fact, I used that set of molar forceps right there in front of you."

"Did ya have to give him somethin'?"

"Absolutely," I said. "Had to knock him out completely."

We worked in silence for a while, Hersey steadying the horse's head as I clipped back the disproportionate molars. Just as we were finishing, Hersey said, "I got a bad tooth in my mouth, been botherin' me for quite a spell. I pulled two of 'em myself awhile back, but I can't reach this one with my pliers. Don't s'pose you could pull it for me? It's all loosened up."

"Good lord, Hersey, have you ever thought about going to a dentist?"

"I never sat in a dentist chair in my life and I ain't about to now."

Clearly Hersey's stubborn streak didn't end with his horses. "Well, here's what I'll do. Let me place the pliers on the tooth for you, and you pull it out."

He nodded. "That'd help. But let's go up t' the house where it's warm."

In the kitchen, Hersey yanked out the bad tooth without batting an eye. Afterward, he fished a bottle of Mr. Boston whiskey out of his oven and poured half a glass, swished it around his mouth for a minute, and swallowed. He swiftly poured another half-glass and downed it in two gulps.

"Gotta be careful 'bout gettin' infection, with all the germs hangin' round nowadays," he said, screwing the cap back on the bottle. "Hell, I forgot. Maybe you'd like a pull, Doc." At that point, I didn't mind if I did. The rotgut seared my throat, but after I shook off the taste, the warmth was welcome.

After I passed the bottle back to him, he said, "Now, about your money, which I ain't got. But I'll tell ya what. Our hens are about through layin' and I'm gonna dress 'em off. How 'bout if I give ya five hens and a bushel of potatoes?"

"Okay, but I'd rather have gasoline if you can spare any."

"Sorry, Doc, I don't have a drop on the place."

"Does your neighbor have any?" I asked.

Hersey shrugged. "Don't know, but the place is all locked up anyway. They've gone off visitin' for a few days."

I nodded and explained to him that I had knocked their clothesline down, and he agreed to go up in the morning and fix it. He fetched the potatoes from his cellar while I warmed up my car. After I placed the bushel into my back seat, I set out on the same icy road in the reverse direction. By sheer luck I made it back to the main highway, my gas gauge needle bumping on empty. This was no night to get stranded in the boondocks. I decided to take a shortcut on an old road that was uninhabited but would lead me into a hamlet about three miles away. The shortcut was icier than the main road but had no steep hills, so I took the odds. Fishtailing wildly, I made it about halfway down the road when the car started a long skid. The snow was deep enough to slow the motion, and the car came to a gentle rest in a ditch, with the driver's side tilted down toward the bottom of the ditch and braced against the opposite bank. Even without looking, I knew that the front and rear wheels on the passenger side weren't touching the ground. I shut off the engine to conserve whatever gas I had and pondered my next move.

It was now about 1:30 in the morning and at least twenty below zero. The wind howled across the field beside me, packing the snow into a dry, felt-like consistency. Though dressed for cold weather, I knew that walking the three or four miles back to Hersey's was too risky. If I bogged down, all bets were off.

Fortunately, the car would still run, as it hadn't tipped the full ninety degrees onto its side. In the rear-view mirror I saw that the passenger-side wheels, which were closest to the road, hung only a foot or so off the ground. I climbed out through the passenger door and pulled down on the passenger door handle. The car rocked, almost enough to right itself—but not quite. As I scouted around for a couple of short logs, the frigid wind cut through my layers of clothing and straight into my bones. I forced myself not to think about what could happen if my plan failed.

I found a small, stout tree and dug my axe out of the trunk.

Being on the road so much, I kept all manner of tools on hand for emergencies. I chopped down the tree and cut it into a short blocky section and a longer one. Using the short piece as a fulcrum, I set it in the ditch alongside the driver's side. I stood in the ditch and started prying with a longer log, rocking the car until it settled back onto four wheels, though it still tilted at a precarious angle. I climbed in and edged down into the driver's seat, hoping that my two hundred pounds wouldn't tip the passenger side into thin air again. I turned the key. The engine started without a hitch.

I looked straight down the ditch and saw that it grew gradually shallower for the next hundred feet until it became level with the road. I hit the accelerator. During the next two or three minutes, a succession of emotions—fear, anxiety, elation, and sheer joy—raced through me as I roared down the ditch like it was a salt flat in Utah. With a final lurch, the car slewed onto the road.

After spinning and skidding for another few miles, I came to the end of my "shortcut" in Searsmont Village. All was dark, save for a single light in Ollie Waters's farmhouse. Ollie was a friend and a client of mine. I banged on the door and waited. At last Ollie shuffled to the door, rubbing sleep from his eyes. "For God's sake, that you, Doc? You look half-froze to death. Come on in and get warm," he said. "Fell asleep watching the late show. Let me put on some coffee."

I stood by the oil stove in Ollie's old farm kitchen and told him about my plight. He shook his head in wonderment. "We'll get some gas into that rig of yours. But son of a gun, I was gonna call you this mornin' anyway. I got two cows off their feed for two or three days now. Wouldn't take a look at 'em while you're here, would ya?"

After we had downed hot coffee and homemade doughnuts, Ollie pulled on his coat and boots and went to the shed to fetch some gasoline and a funnel. We poured it into my tank and then went to the cow barn to take a look at the cows. They stood side by side and shared the same automatic-fill drinking bowl. The scant, hard feces scattered behind them told the rest of the story.

"Ollie, this bowl's drier than a bone," I said. "Looks like the water valve is stuck shut."

I didn't want to embarrass him, but dehydration this severe was life threatening, and obviously he hadn't thought to check the water. "Well," I said, "I'm sure you won't forget again."

He nodded, red-faced, and we settled our accounts. The small charge for my services evened out the cost of the gas. I was reaching for the door handle of my car when Ollie said, "Damn, I almost forgot—Delvin Pickett wanted you to stop over to his place and pinch two bulls when you were out this way sometime." By "pinch," he meant "castrate."

It was three in the morning, but I decided to take care of the job then, rather than scribble a note to myself and try to find it later among the many scraps of paper on the floor of my car. I drove the half-mile to Delvin's place and knocked on his door. A few minutes later he swung open the door, shotgun in hand. I put my hands up.

"Jeez, Doc," he said, putting his gun away. "You give me a fright." He pulled on a heavy coat and led me to a little shed behind the house, and with his help, we pinched his young bulls in short order. (The procedure is a nonsurgical vasectomy, in which the vas deferens is crushed with a pliers-like instrument called an emasculatome. When the animal is only a few months old, it's relatively painless.) I charged him my "when by" rate, which made up for his lost hour of sleep. It was dawn when I pulled back onto the main road and headed for home.

I never did get those fowls from Hersey Clements, as he died several weeks after my call there. I wondered whether he had succumbed to an infection contracted from his pliers, but I never did find out.

13

The Choking Ox

One summer Saturday I had been working exceptionally hard since dawn in order to get home for dinner with friends at seven that evening. It was about four o'clock and I had just left a farm in the hill-town village of Washington, when I got an emergency radio call from my office: Forrest Hovey's ox was choking to death in Liberty, about fifteen miles away via a series of twisting back roads. I shifted the car into overdrive and was making good time when I hit a very sharp curve, the kind Mainers call a "square round turn." The car was shuddering and its tires screeching on the hot summer tar as I rounded the curve and broke onto the straightaway.

There, directly in front of me, were four state police cars parked in my lane. The cruisers were lined up end to end, and the four troopers were leaning over the trunk of the nearest car, chewing the fat. There was nothing to do but swerve violently around them. As I rocketed past, the Staties scattered like quail and dove for the ditch. Luckily nobody was coming at me in the other lane, and I kept on racing down the road. Every few seconds I glanced at the rear-view mirror, expecting to see a trooper in pursuit. Though my car was familiar to all the police in the area, I was pretty sure they hadn't had time to recognize it as they leapt for the ditch.

Arriving at Forrest Hovey's place, I found his one-and-a-half-ton ox, Star, flopped on his back and wedged in the narrow door of the tie-up, straining to breathe and swallow. His midsection was bloated tighter than a drum, which made the poor critter look like a convulsing Goodyear blimp. The fact that he was still alive meant some air was getting past the obstruction in his windpipe. Forrest and three or four of the neighbors were clustered around the ox, and one held a hunting knife. No doubt they were about to relieve him by puncturing the animal's rumen, releasing the trapped methane gas that bloated him. They stopped when they saw me pull in.

"Boy, am I glad to see you, Doc," Forrest said when I ran over, grip in hand. "Star's got a whole potato stuck in his throat, and he's just about done. We were about to stick him but weren't sure where to put the knife. I should a never fed him those potatoes. You think you can save him?"

It was commonplace for farmers to feed cattle potatoes as a part of their daily ration. This easily digestible food was a government-subsidized crop for dairy farmers during the late '40s and early '50s.

"I'll try," I said as I grabbed my trocar, a tubular puncturing instrument. The potato had obviously blocked most of his windpipe and his esophagus, the passageway to the rumen—the largest of his four stomach compartments. The rumen of a mature ox could hold about ten gallons of predigested food, and if air couldn't get in through the esophagus, the food would begin to ferment, producing methane gas, a poison. The trapped methane caused Star's rumen to bloat painfully, and as the gas began to be absorbed into the animal's bloodstream, it could kill him. I quickly inserted the trocar at the highest point of the distended area, puncturing the rumen. Like an emptying balloon, the ox's swollen belly deflated with a loud rush of foul air.

With emergency number one attended to, I turned to emergency number two—the potato stuck in Star's throat. I attempted to snake a stomach tube into his throat to dislodge it. No longer pinned in the doorway, the giant beast tossed me about like a willow in a windstorm. Dodging and weaving, I eventually succeeded in pushing the tube into him, but the potato wouldn't budge.

"Forrest, do you have a garden hose?" I asked.

"Yup, I'll fetch it," he said, hurrying off.

I desperately continued trying to dislodge the potato, twitching the stomach tube this way and that. Before Forrest could return, the ox suddenly stopped thrashing, and his eyes rolled back into his head from lack of oxygen. He crashed down in a heap and started to go into cardiac arrest. I grabbed my jackknife and performed a tracheotomy, cutting a small opening between the second and third tracheal ring to let air into his lungs. It also allowed him to swallow again. Then the neighbors and I jumped up and down on his massive chest until Star started breathing. Before he regained consciousness, I stuck my bare arm into his throat to see if I could reach the potato, to no avail. All I got for my trouble were a couple of badly lacerated fingers from the involuntary grinding of Star's giant molars.

By now Forrest had returned with his garden hose. I lubricated the brass fitting at the end with mineral oil and stuck it down the beast's throat, using a mouth speculum to hold his jaws open. I ground and poked the metal into the potato, and about a minute later I felt the potato break up. I withdrew the hose and reached in to clean out the pieces, using my injured hand to ensure that I would have at least one left after this adventure. The ox was beginning to stir as I raced to suture the tracheotomy, and I managed to get in the last stitch as he tried to right himself. The whole procedure had taken twenty minutes. We all watched in relief—Forrest with a tear in his eye—as the giant critter stumbled to his feet and blew out a great snort of air.

After collecting my gear and giving Forrest's dog a while-you're-here rabies shot and checkup, I patched up my fingers and got back on the road. Well, I thought, I might actually be on time for a dinner party for once. Just then my assistant's voice crackled over my two-way car radio: "Junior Buzzy's Arabian stallion just went through a barbed-wire fence and got a deep cut near the groin. He wants you to come as soon as you can."

"I'll be there in about half an hour," I said, signing off. Junior Buzzy lived on the outskirts of Belfast, and it was on the way home anyway. With any luck, it would be a straightforward sewing job, which I should be able to dispatch quickly using only a local anesthetic and a sedative. It all depended on who Junior had lined up to handle the horse, because he himself was no horseman—in fact, like a surprising number of horse owners, he was scared to death of the animals. Coming from a monied family, Junior kept Arabians mainly to show them around the country, keeping up his status in the process.

I turned into the long driveway of the Buzzy property and parked by the paddock, where Junior and another man stood next to a handsome Arabian with one hind leg held up. I recognized the stallion as Sheik, who was a handful for even an experienced horseman.

When I got out of the car and walked closer to the group, I saw that the second man wasn't exactly standing after all, but swaying. Closer still, I smelled a powerful aroma: whiskey. Junior, three-quarters drunk himself, introduced me to his inebriated friend. "Drew, meet my veterinarian, Dr. Brown," he said, slurring the words a little. "Drew's in the bankin' business, Doc, and owns some fine Arabs. Handles his own show horses." Drew gave me a looped grin, apparently quite pleased with himself.

"Junior," I said, "Come help me carry some things from the car."

Walking to the car, I asked him in a low voice if I could trust Drew to hold Sheik.

"Absolutely," Junior said, a little too loudly. "He's been around horses all his life. We met him and his wife at a horse show in Connecticut a couple of weeks ago."

This answer was not what I was hoping for, but with no alternatives, I started the job. First I injected Sheik with an intravenous sedative, and after a few seconds, bent down to inspect the wound: an ugly foot-long gash began at the end of his scrotum and extended down the inside thigh. I straightened up. Sheik danced skittishly in place, and Drew staggered as he fumbled with the horse's stud shank—a chain bit with a leather strap. "Hooo, boy," he said.

After telling Drew what I was about to attempt, he assured me that there'd be "nooooo pr'blum" holding Sheik steady. Saying a silent prayer, I bent down on one knee under Sheik's bobbing belly and started injecting the edge of the wound with a local anesthetic. No sooner had the needle entered his flesh than the horse shifted his weight onto his front legs. Trapped, I saw the whole thing coming. One of his rear hooves crashed into my rib cage and the other smashed into my left quadricep. I felt the sickening tear of muscle as I toppled over, sucking air. "Oh, Lord," I heard Junior say.

After several seconds, I managed to regain my wind and struggled to my feet. Sheik was missing in action, and Drew was clutching a bare, bloody shoulder near the ragged seam of his shirtsleeve, which was also missing. It looked like Sheik had taken a souvenir before escaping to the safety of the neighbor's lawn.

"Good God, Doc," Junior said. "If he'd hit your head you'd be pushin' up daisies."

"Yeah," I said, wincing as I tested my weight on my left leg. "But while I'm still here, let's catch your horse." Junior nodded. Drew stayed put.

About ten pounds of grain later I got hold of Sheik's stud shank and wrestled him back home. Junior offered lots of encouraging words, but no other help. This time I put Sheik under general anesthesia and everything went smoothly.

I was tying off the last stitch when we heard a loud screech of rubber out on the street and the sudden yelp of a dog. Junior ran off to investigate and came back carrying one of his young beagles.

"Thank God you're here, Doc," he said, placing the whimpering dog on the grass. I stopped what I was doing with Sheik and checked the beagle over. He had been lucky, sustaining only a broken hind leg. After giving the little fellow a painkiller I made a temporary splint out of an old cedar shingle I found lying on the ground, which would do until I could get to the hospital and do a professional job. I then turned my attention back to Sheik.

It was another half-hour before the Arabian was steady on his feet again. I placed the groggy little beagle in the front seat of the car with me and headed to my office. When I got there it was about 8:30, so I called home to report the obvious: I was late, as always.

An X-ray of the little dog's leg revealed a compound fracture, which required surgery to insert a stainless-steel pin and plating. After finishing the operation, I bandaged my swollen leg and the cuts on my fingers. Finally, at almost 10 P.M., I hopped in the car for home.

Our living room was filled with laughter and familiar faces. Then I noticed one couple whom I had never met before. They had come up from New York City for the weekend to visit some mutual friends who were also at the party. After introducing ourselves, the man of the couple said, "I don't imagine you get too much vet work up this way. You ought to come to New York. Our vet down there makes as much as an M.D. and works only thirty hours a week."

"Oh, I don't know," I said, "I manage to get in a full week." My friends, overhearing the exchange, burst into laughter.

A few days later, one of the state troopers stopped by my office and told me he had recognized my speeding car on that wild Saturday afternoon, but he and the others agreed that they had been stupid to park in the road, so they let it pass. He did recommend that I slow down, however, reminding me that I had repeatedly received the same advice from other troopers.

I agreed that slowing down would be a fine idea.

14

Bess

One Sunday afternoon Pete Murphy called, very upset. He and his nine-year-old mare, Bess, had been galloping down a woods trail when Bess had caught a branch in her right eye.

"I can still see a piece of the branch sticking out of the eye, Doc, and she's in hellish pain," said Pete. "Looks like the eye's had it—it's a mess."

"Take a level teaspoonful of salt, Pete, and put it into a glass of warm water. Stir it and keep washing out the wound. I'll get over just as soon as I can."

I hung up, heartsick.

To know Bess was to love her. She was a proud white mare, a well-muscled horse who drew admiring glances from many a horseman. It was my good fortune to have been her vet since I brought her into the world. Over the years, she had won her share of blue ribbons at fairs and shows. Spunky but gentle, Bess could barrel race with the best of them, and she relished every minute of it. She had taught dozens of children how to ride and yet she was a fierce competitor in shows and events. One of the best all-around horses I'd ever met, Bess loved everybody, and the feeling was mutual. So I cut a swath through the heavy summer traffic on Route 1, covering forty miles in half an hour, knowing that every minute I saved would spare Bess that much suffering.

When I got there, her eye was even worse than I had imagined. Pete was right: it was completely destroyed. It would definitely need to be removed as soon as possible to prevent more pain and ward off the possibility of an infection ascending into the brain—that much was clear to me medically, and to Pete from a humane standpoint.

The only question was how Bess would function with one eye. Would she adjust or become frustrated and anxious?

Pete turned to me, his throat choked with emotion. "Do you think it would be fair to her to remove it, Doc, or should we put her to sleep?"

"Well, Pete, I've unfortunately had a lot of experience with this question. I've removed irreparable eyes from horses, dogs, and cats for many reasons, and I can't recall a single patient who didn't adjust to the handicap."

My answer wasn't based on the fact that I couldn't bear to put Bess down. I operated on the general principle that we wouldn't consider euthanizing people with damaged limbs or eyes, so why should an animal be any different?

Pete's face relaxed. "Okay, Doc. Let's go ahead."

Using a general anesthetic, as well as a local one for the eye, we proceeded to do the job on the lawn. A few neighbors, seeing my familiar car in the yard, came over, worried that something had happened to Bess. A handful of them lingered to observe the operation, although some drifted off when the going got gory; removing an eye is a hard thing to watch.

The surgery went smoothly, and Bess was on her feet an hour and a half later, eating apples out of Pete's hand. As I was washing up my surgical gear I noticed a small boy walking toward us, leading a mongrel dog on a piece of baling twine. When he got closer I saw that tears were rolling down his cheeks. He looked to be about six or seven years old.

"Are you Doctor Brown?" he asked.

"Yes, I am," I replied. "What can I do for you, son?"

"This is my dog, Rover. Can you look at him while you're here? My mother's boyfriend says he's gonna shoot him this afternoon cause Rover's got some bunches growing on him. We took him to a vet'nary yesterday and he said it would cost fifty dollars to take off the bunches. But we don't have any money." At this point, the little boy broke into sobs.

I squatted down and patted Rover on the head. "Where's your father, son?"

"He's gone to heaven," the boy said. "He drowned in the ocean two years ago."

"Was he a fisherman?"

The boy nodded.

"Do you have a name?"

"George Thomas."

"How many brothers and sisters do you have?"

"Four brothers and three sisters."

"I see. And how old is Rover here?"

"He's eight years old, a year older'n me."

"Well, let's take a look at him."

I examined the bunches on Rover's head while he lapped my face. After feeling them, I was certain that the bunches were lipomas, or benign fatty tumors, easy enough to remove surgically.

"Are they cancer? My mother's boyfriend said they were and we'd all catch it."

"Well, neither is true, George. These particular bunches are not cancerous, and even if they were, you couldn't catch it from Rover."

"Will you fix him, Doctor Brown? People say you like animals."

I laughed. "Yes, that's true. I'll try to fix Rover, but first let's go talk to your mother."

After telling Pete about how to care for Bess, Rover and George hopped into my car and we headed for his house about three miles away. As I drove into the yard the family's poverty was evident. No electric wires ran to the house, and when the young mother answered my knock at the door, I saw that stress had etched deep lines in her face. The kids showed the same signs: weary eyes peered from colorless skin, and their clothes were tattered. George and his mother let me in the kitchen door. I glanced about and saw no fixtures to indicate running water.

The mother's boyfriend sat at the table with an open can of beer in front of him. Feeling his beer, he immediately took center stage. I explained Rover's problem was not serious and that I'd take him home with me and remove the growths free of charge. Mrs. Thomas, glancing nervously at her boyfriend, stayed silent. The boyfriend, swilling from his can and swaggering around the tiny living room, announced that the dog was nothing but a nuisance—just another mouth to feed. "Besides," he concluded, "he's only a mongrel. If he was a purebred it'd be different. And I tell ya, I ain't about to catch cancer from no dog."

All of the children were now sobbing openly as this genius continued to rant. Finally I interrupted him.

"You know, mister, you're not a purebred, either," I said. "Neither am I. People are all mongrels just like Rover, and furthermore, nobody is going to catch cancer from this dog. First because he doesn't have cancer, and second because it's not possible to catch cancer that way."

My little speech gave Mrs. Thomas a chance to assert herself. "Well, Roger, if the vet'nary is willing to fix Rover for free, then let him do it. After all, he belongs to the kids, not you."

Roger sucked down the last of the beer in the can, crumpled it in one hand and dismissed her with a wave. "Go on, then. See if I care."

Assuring the children I'd return him the next evening, I got into the car with Rover and headed to Belfast, sharing a half-pound of liverwurst and crackers on the way. Later that evening, I returned to the animal hospital after my farm calls and removed the fatty tumors handily. After the anesthesia had worn off, I treated Rover to a can of chicken with a bit of raw hamburger mixed in. For dessert we shared some vanilla ice cream.

The next evening Rover and I made some more farm calls together on the way back to his home. George and his brothers and sisters had apparently been on the lookout, and they started running down the roadside when they spotted my car. Rover pressed his paws to the window and began to bark, his tail slapping the car seat. Once Rover hit the yard there was an explosion of joy. He ran wildly from one child to the next, lapping their faces. Once the initial excitement died down I brought out a case of ice cream bars that I had bought along the way and we all celebrated the homecoming. I was relieved to see no sign of the know-it-all. Mrs. Thomas came out of the house and gave Rover a kiss herself. "You know, Doctor Brown, these kids haven't eaten a bite since Rover went off with you," she said. "They didn't believe you were going to bring him back."

Several weeks later I went through the mail and found a small, clumsily wrapped package. Inside was a whittled wooden whistle with a note. "You know we don't have any money to pay you for fixing our Rover, so I made this whistle out of a green willow branch.

Rover is doing good. Hope you like the whistle. Your friend George Thomas."

I liked the whistle very much indeed.

Bess, I'm happy to report, compensated very well for the loss of her eye and went on to teach countless more children to ride. And she astonished Pete by barrel racing with the best of them, living to a ripe old age.

Rover lived to be seventeen, a very old age for a dog. In his later years he contracted severe arthritis, which rendered him increasingly crippled and in a great deal of pain. Finally I suggested to George, who was by then sixteen, that we should put Rover out of his misery. So one day George brought him into the animal hospital. As he held his dog, I slowly injected Rover with an overdose of anesthesia and we watched him leave this world in a peaceful state. Through his tears, George thanked me for giving Rover nine extra years of life. With that, he left to bury him in the backyard.

15

Jeptha Goggins's Emergencies

Sometime deep in a November night, the phone rang. I groped for it before squinting at the clock..

"That you, Doc?" I recognized the voice of Jeptha Goggins, a farmer in Waldo.

"Yeah, Jeptha, it's me. How's everything with you tonight?"

"Hell, it ain't night out here, Doc—it's three in the mornin'."

"Fair enough. So what's up, Jep?"

"You know that cow you call Ada? Well, she's got her yearly problem, only worse. On top of the milk fever she's cast her wethers. When can you come out?"

"I'll be right along," I said, sliding my feet to the floor. "How's the weather out your way?"

"It's been sleeting here most of the night," he replied, as though it were obvious. "Can't hardly stand up, the dooryard's so damn slippery. If ya can't make it over Devil Hill you'd better go around the mountain and come up the back way. It ain't quite so steep, but I dunno if you can make it that way either."

"Well," I said, "I'll give it a try, Jep."

With that I hung up and got dressed for the twenty-five-mile journey to Jep's hilltop farm. I had no idea then that of the countless times I had traveled over slippery roads in my hundreds of thousands of miles of driving, this trip was to become one of the most memorable. I set out, confident that my car was as ready for the weather as the rear-wheel-drive technology of the time allowed. I was equipped with barium-studded tires and positraction, a gear mechanism that allowed the wheels to better grip the road. Even so, the ice gave me a few scares and I got a little concerned, knowing that the real test lay ahead—making it up the steep hill to Jeptha's farm.

As if the main road weren't bad enough, the five miles of narrow side road into Jep's were much more formidable. A light rain began to fall, covering the icy road with water. With the rear end sliding right and left, I barely made it up to the brow of the first hill. In front of me was a short, V-shaped valley, beyond which rose the sharp crest of Devil Hill. Steadying myself, I started down into the valley, letting the vehicle gather speed to make it up the long, steep hill. Everything was going my way as I streaked down the slope, crossed the valley floor, and continued speeding up Devil Hill. Then, within a hundred yards of the peak, the car's rear end started sliding from side to side, slowing my speed to a crawl. But I was still inching ahead and optimistic that I would make it to the crest.

Then my eyes did a double-take. Coming over the brow of Devil Hill was a seventy-foot tractor-trailer rig stacked with crates of live chickens. The driver must have just loaded up at a poultry farm just a short distance from Jep's. Frantic, I tried to back down the hill I had just conquered, but about a third of the way down the car started sliding uncontrollably and came to a stop across the narrow road. My vehicle blocked the road completely as the truck began its descent.

By now the truck driver must have shared my horror, knowing he wouldn't be able to stop. I scrambled out of the car, fell flat on my face, and slid down the hill. Still on my belly, I managed to dig my fingernails into the ice and steer myself into the ditch. Then the crash of metal and glass filled my ears. I watched helplessly, only two feet away from the passing wheels of the giant rig as it pushed my vehicle into the ditch on the opposite side of the road. The big rig managed to stay on the road and continued to the top of the next rise before stopping. I heard hundreds of chickens squawking. The driver leapt out of the cab and headed back toward the wreck, slipping and falling several times. I had regained my footing by this time and waved, signaling I was okay. When he made it to where I stood, he was obviously shocked to see me not only alive, but uninjured. He started apologizing, telling me he had had no choice but to hit me. I told him I understood, and we exchanged insurance information so he could get on his way.

"By God," he said in parting, "we almost had Maine's biggest chicken barbecue."

Jep had heard the crash and came down as fast as his seventy-five-year-old legs would carry him, sticking to the shoulders for traction. After he caught his breath, he allowed he'd get his big farm tractor with the heavy-duty Canadian chains and try to pull my car out of the ditch.

"Not now, Jep," I said. "It's out of the way. Let's tend to that cow first." Grabbing my grips out of the trunk—fortunately it wasn't smashed in—I accompanied him to the barn. My body was still charged with adrenaline, and I scarcely noticed my bloody fingers or felt the relief of surviving the accident.

Ada was in bad shape. Milk fever had left her paralyzed and nearly comatose. She was stretched out flat on her side, her posterior and everted uterus in the manure gutter. I quickly gave her a 500-cc bottle of calcium solution and my damaged fingers started coming back to life. Feeling jolts of pain, I looked at my hands and saw for the first time that every one of my fingernails had been broken off below the quick as I'd slid down the icy road. My hands were shaking from the trauma. Not now, I thought. Just focus on the cow.

The calcium solution roused Ada to full consciousness and she lifted her head off the floor. Now if she would only get up on her feet, it would be a lot easier to squeeze her uterus back into her body. In fact, with her hind end hanging in the gutter, it would be nearly impossible to work it back into place, even if I stripped to the waist in the icy barn and lay in the gutter beside her. This I knew from experience. So after removing the placenta, which was still clinging to the uterus, I sat with Jep and waited.

After thirty minutes she still hadn't gotten up, so I tried my electric prod on her to see if any strength had returned to her rear leg muscles. But she didn't seem to feel it. Even after we replaced the batteries, it had about as much effect as a mosquito biting an elephant. We decided to shock her by tapping into the current in the bare "training wire" that ran above the shoulders of the cows. (Electrified wires train the cows to back up while defecating and urinating, so that they hit the gutter rather than their stalls.) We stripped away the insulation from both ends of the phone wire and planned to touch one end to the overhead training wire and graze Ada's rump with the loose end. The current might be enough to jolt her to her feet.

Jep's two grown boys arrived in the barn to start their 5 A.M. chores, and Jep beckoned one of them, Eldon, to lend a hand. Being in his late twenties and built like an ox, he was a definite asset.

"Now, Jep," I instructed, "you hook that piece of phone wire onto the training wire. When I tell you, touch the loose end to her rump. If she tries to get up, I'll be on the tail end, lifting. Eldon, you grab her flank." They nodded.

Poor old Jep, slow from arthritis and a little forgetful, gripped a steel post to steady himself. With his bare hand on the exposed end of the phone wire, he touched it to the training wire overhead.

"Dad!" Eldon shouted. Too late. The shock dropped Jep to his knees, and we thought we had lost him. But luckily the wire had slipped out of his hand when he fell, and he was only dazed.

"Gorry," he said as Eldon and I helped him to his feet. "What'd I do, anyway?"

We parked Jep on an old milking stool to recover. He watched as I made a loop out of one of the exposed ends of the wire and secured it to the overhead training wire.

"Now, Jep," I said, "we don't want to lose you, so you hang onto the insulated part when you touch her with the other end of this thing. And don't touch the pole. Okay?"

"Ayuh," Jep nodded. "Reckon I forgot. I'll do better this time."

After positioning ourselves again to assist Ada, I told Jep to graze her rump lightly with the wire. He stepped around the downed cow, stumbled on the edge of the gutter, and dropped the end of the wire into one of his big barn boots, where it zapped his leg. Down he went, moaning. Eldon dashed to his side and yanked the wire out of his dad's boot, and we helped him to his feet again.

After a moment of confusion, Jep muttered, "Geez. Guess I'd better watch what I'm doin'. Let me set a spell and get a chew of tobacco. Then I'll see iffen I can't do it right."

Eldon went off to see how his brother was doing with the chores while I kept an eye on his father. When Eldon came back, we tried our plan again. This time it worked. With a lot of pushing, Ada struggled to her feet, a forty-pound uterus flopping around her rear hocks. She shifted her weight woozily, her legs still unsteady as the muscles came out of paralysis. Jep and Eldon murmured encouragement to her as I worked fast to take advantage of her dulled senses. I gave her a mild spinal block to prevent straining and replaced the uterus. Thin cellophane gloves protected my fingers a little, but the effort was pretty painful.

As we were finishing, Jeptha's wife, Rose, came running into the barn. "Eldon, come quick," she cried, gasping for breath. "Your wife's in heavy labor. Ain't no way you're gonna get her twenty miles to the hospital on them icy roads. You're gonna have to give me a hand."

Eldon bolted for the house, leaving his mother to trot behind. Jep started after them while I washed up and packed my instruments.

Five minutes later Eldon returned to the barn, distraught. "Looks like the baby's coming right off," he said. "Lilly's in awful pain. Ma's delivered a lot of babies, but since you're here, Doc, we'd be much obliged if you'd help out, seeing as it's Lilly's first baby."

"Sure, Eldon. Why don't you help your father with the tractor and see if you can get my car out of the ditch. I'll go help your mother with Lilly."

Eldon nodded and dashed off. I found Lilly already halfway through the delivery, with Rose lending a competent hand. "God, Doc, we never meant to have one of those natural childbirths, but here we are," Rose managed to joke. Lilly didn't share in the humor, but she pushed like a trouper, and twenty minutes later she delivered a healthy, seven-pound boy.

While washing my hands in the kitchen, I looked out the window to see Jep and Eldon towing my car into the yard. I went out and told Eldon he'd better come in and meet his new son. He stared at me for a second before dropping into a dead faint. I grabbed a wet dishtowel and placed it on his forehead while Jep elevated his feet. In a few minutes Eldon came around, and we got him into the house. Jep and Eldon stood at Lilly's bedside, admiring the red-faced infant.

"Fine lookin' boy we got, Doc. Thanks for your help." He wiped a tear from his eye and pumped my hand. I flinched, but managed a quick smile. It was his first grandchild.

"No problem, Jep. Nature and Lilly did most of it, with help from Rose. How's my car looking?"

"Well, the right side of her is all stove to hell, but the underpinning looks all right. We had to pry the fender off the tire 'fore we could move it." He shook his head. "Them Canadian chains are really great on ice, but even they were spinnin' all the way. You were mighty lucky, Doc."

I nodded. The whole thing still hadn't had a chance to sink in. I started the car, a miracle in itself, and gave it a test drive in Jep's yard. It creaked but seemed basically okay. At some point during our escapades, the sand truck had come through, and I waved to Jeptha and headed back out on the road. My trip back to Belfast, to my great relief, was a lot easier than the ride out.

Jep called a few days later to report several new vet problems, none of them urgent. He asked me to stop by the next time I was out his way. Before hanging up he said, "By the way, Doc, Ada's doin' good, and that new grandbaby of mine is right smart. They named him after me. Bet you can't guess what Lilly and Eldon nicknamed him."

"Jep II?"

"Nope. They've taken to callin' him 'Doc.'"

16

Swimming with Ringo

At about 7:30 one April morning, I was finishing up some routine surgeries in my hospital when the phone rang. The hospital manager was still hosing down the dog runs in the back, so I picked up.

"How much does it cost to fix a stud horse?" the male caller asked. He didn't identify himself, and I didn't recognize the voice.

"Generally I charge thirty-five dollars, barring anything unusual," I answered.

"Well, when do you suppose you could do it?"

"When the weather warms up a bit in May," I answered. "Who's calling?"

"Basil Trent. I'm in Brooks at the old Benny farm. You haven't been out here since I bought the place. They used to have cows here, and I've turned it into a chicken farm, but I've got a couple of horses, too."

"Sure, I know the place. You want to do it out there, or can you bring the horse in here?"

"I don't have no trailer, so you'll have to come out here."

"Okay. I'll call you some warm day in the next few weeks and we'll take care of it."

Several weeks later I was making some farm calls in Brooks, and as it was warm enough for outdoor surgery, I decided to run up to Trent's farm and castrate his stallion. I gave him a call from the local store to make sure he was home, and he said to come right over.

The road into the farm was almost straight up for a good mile, and the town road grader happened to be working on it that day, making it temporarily impassable for the last quarter-mile. Not wanting to wait, I parked on one side of the road, grabbed my grips, and hoofed it the rest of the way to the farm. Walking toward the house, I surveyed the lay of the land to pick a spot for the open-air surgery. A flat grassy area on the front lawn looked just right.

I didn't see anyone around, so I knocked on the front door of the house. Basil Trent, a lanky man of about fifty, stepped out.

"Hi," I said, "I'm Brad Brown."

"Oh, hi, Doc. Didn't hear ya knockin'. Where's your automobile?"

"Left it down the road a quarter of a mile or so. They've got the road all torn up on this end, so I decided I'd walk to save time."

After I explained exactly what we were about to do and where, Basil asked me what he should do with the "stones," or testicles, once they had been removed from his stallion.

"Well," I said, surprised by the question, "usually people bury them, but if there's a dog around they usually gulp 'em right down. Dogs seem to relish them."

Basil frowned. "I don't think I'd let my dog have 'em, even though we don't have no relish. I was talkin' to Weasel Prouty the other day and he said 'round the full of the moon the stud'd bleed to death iffen I didn't put the stones on the highest beam in the barn."

This was a new one. "Did he say why?" I asked.

"All the horse's blood would try to get back into his stones. But if I was to put 'em on a high place, the blood can't run uphill."

"Well, that's just not true," I said. "The only reason he'd bleed to death would be if he were a hemophiliac."

"Well," Basil snorted, "that won't happen, 'cause he only likes females."

"That's not—" I started to say. "Oh, well, forget it. We'll check his blood for clotting before cutting, and you can do as you please with his testicles." Basil nodded.

The stallion, Ringo, was seven years old and well built. And I saw immediately that he had never been trained. In fact, Basil told me he had raised Ringo from a colt and hadn't been able to get a halter or any other rope on him since he was a year old. Now, six years older and weighing twelve hundred pounds, Ringo presented us with a formidable task.

"Basil, may I ask why you haven't put a halter on him for the last six years?"

"Well, it weren't my idea to keep a horse. My wife, Edna, wanted him. She's never rode a horse or anything, she just likes horses. You know how it is—she fell in love with the little fella when he was born and wouldn't let me sell him. 'Course, as he got bigger and started biting her, she got a mite skittish. Brings him carrots and a sugar lump every day, but she's still scared to death of him. He's gotten so studdy, he's staving his stall to pieces and bitin' at her. She figures cuttin' will calm him down. She bought a new halter awhile back. Thought we could put it on him while he's asleep."

"Right," I said, "but first we have to get him to sleep."

As if in response, Ringo snorted and kicked the stall door with his hind feet.

Readying two lassos, I told Basil that he would handle one rope and I the other. After ten or fifteen tries, we managed to get both ropes around the horse's neck. Ringo wasn't happy about that, which he communicated by rearing, striking, biting, and generally thrashing around his stall.

Edna heard the commotion and stormed into the barn. "What are you two doin' to my horse?" she demanded.

After we explained, she insisted that we didn't need the ropes. "Just give him a little sugar and he'll stand still for his needle," she said.

Basil shot me a long-suffering look. We relaxed our grip on the ropes for a moment as Edna pulled a sugar lump out of her apron pocket.

"Don't do that," Basil warned. "You're gonna get hurt."

"Shush," she said, clucking to draw the stallion's attention. About a minute passed before Ringo calmed down enough to notice the sugar in her outstretched hand. He darted down, took the sugar, and gave her arm a vicious bite.

Edna shrieked. "Damn it, Basil," she said, backing away and clutching her arm. "He never used to be that way. You've upset him."

"Huh," Basil muttered. "And what's that on your other arm? I suppose he was upset the other day, too?"

Edna sputtered and stalked back to the house. Basil and I returned to the job at hand.

"Okay, Basil, we're going out into the yard with this fella. He's gonna be hard to handle, so whatever you do, don't give him any slack, and try to keep him between us. I'll open the stall door when you're ready."

"Ready," Basil said.

From that point on it was a three-ring circus as we tried to aim Ringo in the general direction of the lawn. Ringo bucked and reared up on his hind legs, dragging us wildly about. After Ringo tore through the yard, throwing gravel in every direction, the lawn looked like a jigsaw puzzle with a lot of missing pieces

"Sure you don't want to put on a rodeo?" I said. "They don't make bucking broncos any better than this."

Basil managed a smile. For someone with no horse experience, he was fearless and quick to respond to Ringo's split-second moves. Gradually, we were wearing the stallion down.

After about ten minutes of wrangling we succeeded in leading him to the flat spot on the lawn for his operation. I started approaching him slowly, gradually shortening the rope between us, talking to him all the while. Basil still held onto his rope, keeping a bit more distance.

Ringo quieted, but he was only biding his time. When I got within striking range, he reared up and lashed out at me with his front feet. I dodged him and slid back to the end of my rope. It was all I could do to hang on as he dragged Basil and me about three hundred yards across the lawn and into the farm pond behind the barn.

It turned out that Ringo and I could swim, but Basil couldn't, which he announced matter-of-factly as he slid heel-first down the steep, muddy bank into the water. He let go of the rope and began dog-paddling like mad, trying to gain a purchase on the slippery bank. He began to sputter.

Alarmed, I let go of my rope and kicked off my rubber boots, which had filled with water and were dragging me to the bottom. I dug my feet into the mud and hauled Basil, coughing and gagging, up over the bank.

"You okay?" I asked.

He sat down and nodded, coughing. I saw that Ringo, in his panic, had swum to the middle of the pond, our ropes trailing behind him. It was a good-sized body of water as farm ponds go, covering at least an acre and deep enough that I hadn't been able to touch bottom. If Basil had been unable to claw his way up the steep, slippery bank, Ringo wouldn't stand a chance of getting a foothold.

I stripped down to my jockey shorts and dove into the pond. The ice had gone out only a few weeks earlier, and without my layers of clothing the water hit my skin like a sleet storm. After catching my breath from the shock, I swam over to Ringo. He was gasping with the exertion of paddling with all of his considerable might. I slung one arm around his neck and grasped one of the ropes. It occurred to me that Ringo stood a good chance of drowning. I hollered to Basil and asked if he were okay.

"Yeah," he said. His coughing had subsided.

"Go get your pickup and bring it in as close to the edge of the pond as possible. We have to haul this fella out of here."

Being a fairly strong swimmer was my only asset now, as I slipped underwater to loop the rope around Ringo's belly. Still underwater, I slid the loop down over his rump and tugged it snug, leaving his legs free to tread water. One of the neck ropes still remained in place. I surfaced, clutching the ends of both ropes, planning to throw them to Basil, who had pulled his pickup close to the bank. I started swimming slowly toward the truck, leading Ringo gently by the ropes. With any luck, he wouldn't catch on that I was in charge. He was straining to keep his head above water, and we couldn't afford to panic him.

As we neared the bank, I tossed the neck rope to Basil so we could steer Ringo to the spot with the most gradual slope. Treading water, I explained my plan to Basil. I floated on my back for a minute to rest, then sucked in a big breath and dove underwater. I worked the rope on the horse's rump down to his constantly moving rear legs and tied it just above the hocks.

With his legs tied, Ringo was about to go down, so we had to act now to save him, and everything had to go exactly right. I popped up to the surface with the end of the leg rope and threw it to Basil, who quickly tied it to the rear bumper. Thank God the rope was nylon, and it held the weight of the big horse as we hauled him, feet-first, out of the water. I stayed in the pond, holding his head up, but the poor beast still inhaled copious amounts of water. The pickup was a four-wheel drive, and Basil handled the job like a pro, giving it just enough gas to drag Ringo to safety without hurting him. When the horse's rear half reached the lawn, I gave Basil a sign to shut off the pickup. Ringo's head and forelegs were well out of the water but still draped over the bank, a good position to drain the water out of his lungs.

"Basil, before we pull him all the way up, let's let him catch his breath. Then I'm going to give him a shot of anesthesia right here in the vein of his rear leg. Otherwise he's going to be spoiling for another fight in a few minutes."

Basil nodded. We left Ringo's rear legs tied to the truck, and with him sprawled like a punch-drunk fighter, I crawled up the bank, went back to the barn for my grip, and trotted back to the pond. In this short time the horse had revived considerably, and I injected the drug before Ringo knew what had hit him. In seconds he was snoring.

Basil slowly hauled the unconscious horse to the level lawn. Still in my jockey shorts, I operated on Ringo as he lay stretched out in the warm sun. Basil took the opportunity to fasten the brand-new halter on him. I had just enough time to put my clothes back on before he awakened, and his mood was just as feisty as it had been before his swimming lesson. By the time we got him back into his stall, the yard was in even worse shape than before, and Basil and I were done in.

As I was picking up my equipment, Basil disappeared into the garage. He soon returned with a ladder. "While you're here, Doc," he said, "would you hold onto this so's I can climb up on the roof of the barn with them stones? After all we been through, I sure as hell don't want him kickin' the bucket."

Knowing Basil would not be deterred, I agreed to hold the ladder. "Watch your step," I cautioned. The aluminum roof was slick, steeply pitched, and about twenty-five feet from the ground.

Putting the "stones" in a plastic grocery bag, Basil scrambled up the ladder and onto the roof. I watched with alarm as he slipped a few times on his way up to the peak. Straddling the ridge beam, he reached into his bag and as he bent over to place the stones on the ridge, he lost his balance and disappeared over the other side. I heard him scream. Racing around to the back of the barn, I found him flat on his back in Edna's freshly planted flower garden. About eight inches of compost had cushioned his fall. Basil was conscious but dazed, and I quickly ascertained that he didn't have a serious head or spinal cord injury. He couldn't speak, but only because the wind had been knocked out of him.

I was starting to examine him for other signs of injury when Edna, who apparently had glanced out the kitchen window just after he had fallen, came running across the yard.

"Basil, damn it, get out of that flower garden!" she shouted. When she got closer and saw that he wasn't moving, she looked puzzled. "What on earth have you guys been up to?" she asked.

I explained, and by the time I was finished, Basil regained enough breath to speak. "My leg," he gasped, pointing to his right leg.

Pulling up his pant leg, I saw immediately that the tibia, or shin bone, was badly broken. Within twenty minutes I had fastened a temporary splint on it. Edna promised to drive him to the nearest hospital right away. Later she called to tell me that the X-ray had revealed three fractures of the tibia.

Several weeks later I chanced to meet Basil hopping on crutches into the Brooks country store. "How be ya, Doc?" he said, grinning.

"Fine, Basil. How's the leg?"

He shook his head. "Mighty slow. But Ringo's all healed up. Wouldn't know we'd a even cut him. He's just as full of it as ever." He paused a moment, looking down at the floor. "I tell ya somethin', Doc. That episode just about cost me my life, between nearly drownin' and then fallin' off the roof. And after all that I dropped the damned stones when I fell. Edna saw the dogs scarfin' 'em up, just like you said."

"Well, I'm glad Ringo didn't bleed to death," I said.

"Yeah, about that. I seen Weasel Prouty a spell ago and told him he was a damned fool. You know what he said?"

I shook my head.

"He said Ringo wouldn't start bleedin' until the full of the moon. 'Wait and see,' he says. S'pose there's any truth to that, Doc?"

"None whatsoever," I assured him. "You know, I was wondering. How come Weasel told you to put them in the barn rather than someplace outdoors, like in a tree?"

Basil shook his head as if this was common sense. "Weasel allowed that if I did that, a squirrel would get 'em, 'cause they collect all kinds of nuts."

I had nothing more to say on that subject. Later that summer, Ringo, having lived through several full moons, was sent away for training. He returned with his high spirits intact, but with a lot more manners and respect for humans. And I'll be darned if Edna didn't learn to ride him.

17

The Commune

In the 1960s Maine and other rural New England states saw an influx of "back-to-the-landers"—most of them young people in their early twenties who were seeking an alternative to the "urban rat race" in which they saw their parents engaged. Some of them were self-described hippies; but some of them were simply romantics nostalgic for a simpler way of life in which they could see where their food came from, experience satisfying physical labor, and unplug from consumer capitalism, which they saw as morally and spiritually destructive. Having been raised in cities and middle-class suburbs where farming was a distant memory, they didn't have the slightest idea what they were signing up for—much less any farming skills. They also tended to idealize rural areas—the more isolated, the better.

One day six of these young people showed up in the waiting room of my small-animal hospital. My office manager told me that they wanted to ask me a question.

As it was a busy day, I wasn't able to see them for almost two hours. When their turn finally came, three men and three women, all in their early to mid-twenties, filed into my examining room. Both the men and the women wore their hair down their backs, and the women wore long, India-print skirts. I asked where they were living and from their description, recognized the old Woodbury place in Knox. The last time I'd been there it was a small, rundown farm. A few more friendly questions revealed that none of the young people had had any farming experience whatsoever.

One of the young men, who wore a red bandanna headband, spoke up in a businesslike way. He told me that the group recently had acquired five yearling dairy heifers from one of their neighbors and a yearling bull from another neighbor.

"I see," I said. "Do you plan to run a dairy farm?"

"Oh, no," he said, sounding a little shocked. "We got them to help clear the fields of brush and weeds. Someone told us that we should use sheep, but we couldn't find any in the area, so we thought that cows could do it just as well."

It was my turned to be shocked, as cattle are primarily grass eaters. But I was drawn to a more pressing question: "Has there been any sexual activity between the bull and the young heifers?"

"Yes," said one of the young women, "nearly every day. But our neighbor from New Jersey, who comes up here summers, says they can't get pregnant until they're two years old. Is that right, Doctor Brown?"

I shook my head. "No, cattle have been known to conceive at nine months of age. And it's possible that your bull is also sexually mature. In that case, you may find yourselves with five pregnant heifers."

The group members exchanged nervous glances. "The problem is," I continued, "they aren't capable of bearing young at this age, and it would probably prove fatal if they're allowed to go to term. Normally they aren't bred until they're at least fifteen months old."

At this point they huddled for a whispered conference. Then the young man with the headband faced me and cleared his throat. "We'd like to have you come out and take a look at the cows and see if they're pregnant when you're in our area," he said. "We understand you get out our way sometimes."

I nodded. "Will do," I said. "Oh, by the way, what did you want to ask me?"

They paused on their way out the door. "Oh, right, thanks," said the fellow with the headband. "Do you know anybody that might want some puppies? They're shepherd-collie mix, real cute."

I told them I'd ask around. A week later I was in the area on a sunny June day and made a point to go up to the old Woodbury farm and see what was going on. As I drove up the long dooryard I was greeted by the sight of eight or nine people engaged in plowing—most unconventionally, I might add. Eight young men were tied waist to waist, forming a straight line. They had hitched themselves to an old-fashioned, single-bottom plow. Another male was attached to each handle of the plow, trying to hold the blade into the ground as the line of plowmen pulled mightily—uphill—trying to turn over sod that hadn't been touched for twenty years.

After parking I walked up to the plow gang. "May I ask why you're plowing uphill?"

One of the fellows answered, "Well, it's a right-handed plow, and this seems to be the only way we can turn the ground over in this corner of the field."

I gently suggested that if they started on the upper left side of their plot and plowed downhill, they would end up with the same result. At this point he thanked me, which gave me the courage to add that they might want to consider plowing across the hill rather than up and down, as the sideways furrows would capture more water and prevent erosion. The young fellow nodded his thanks, looking sheepish. He signaled the others to stop, and they gathered for a huddle, much like the folks who had visited my office.

By now others had seen my car, and several people approached. When I introduced myself, someone said, "Oh, you want the animal-care committee."

Just then, the same six people who had come to my office filed out of the house and escorted me to the dilapidated barn cellar, where they kept their heifers and bull. It was immediately obvious that they had been bilked: what they had been sold as yearlings actually ranged in age from about eight to eleven months. I explained that it was urgent to check them internally to see if they were pregnant, and they gave me the go-ahead.

Haltering the heifers, I conducted rectal exams and found that they were all with calf. I convinced most members of the animal-care committee that the only humane thing to do was to induce abortions and to neuter the bull. But the young man who had been wearing the red bandanna in my office interjected. "You know the rules," he said to the rest of the committee. "This is something that everybody needs to vote on."

"Tom's right," one of the women said.

We went back to the dooryard, which contained an empty gallon drum with a tire iron resting on top. Tom picked up the iron and started beating on the drum. The plow team dropped their harnesses, and people began appearing from all corners of the property. Within a few minutes, about thirty-five commune members had assembled, and Tom the animal-care leader presented the situation, with occasional help from me. Despite my reservations about the efficiency of this particular farming operation, it was gratifying to see that democracy was still working. The large majority voted in favor of aborting the young heifers, and the group started to drift apart.

I stood up. "What about the bull?" I asked.

The young people stopped drifting, but no one spoke. "Well, maybe it will help if you tell me what you plan to do with him in the future?" I offered. "Are you going to eat him? Sell him? Breed the heifers when they come of age?"

A general hubbub broke out, and Tom stepped forward again. "No, no, and no, Doctor Brown," he said. "We don't intend to kill this animal or sell him. And we don't believe in depriving him of his sexual rights, either. He's become a member of our family."

"Okay," I said, "this leaves only one solution: why don't you allow me to perform a vasectomy? That would allow him and the cows their sexual freedom without the threat of pregnancy."

I could tell from their faces that they bought it, but the plan didn't receive official approval until I had fielded countless questions. Afterward, I returned with the animal-care group and we set about our tasks. To abort the heifers I entered their rectums with my left hand and picked up the ovary, manually removing the corpus luteum, which means "yellow body" in Latin. It secretes the hormone progesterone, which prepares the uterus for pregnancy. Then I gave them each an injection of the hormone estrogen to assist the abortion.

Neutering the bull was relatively easy because he was so young. The vasectomy—tying off the two vas deferens, the tubes that carry sperm into the seminal fluid—is the same as it is in humans, and the animal-care committee members looked on, fascinated. After the surgery I suggested that Mr. Bull be kept separate from his lady friends for at least six weeks. While I washed up, the young people peppered me with more questions about animal husbandry and veterinary medicine. As I walked to the car with my grip, a young woman approached me.

"Excuse me, Doctor Brown, but would you come into the house with me? I want to show you something."

I followed her into the decrepit old farmhouse, which the commune members had gutted to create large, open living quarters. She beckoned me up a flight of stairs into a second-story bedroom. I was nervous, having heard many rumors about the promiscuity of hippies, and we had no sooner entered the room than she pulled off her sweatshirt. She wore nothing underneath.

"Uh, Miss," I started to say, averting my eyes.

"Are these flea bites?" she interjected. I turned my eyes back to her and saw that her torso and breasts were speckled with sores about the same size as flea or mosquito bites, but each was a tiny blister.

"No," I said, "they're called vesicular sores, but that's just a description, like the word 'blister.' Any number of things can cause them, and to figure that out, you need to see a physician."

"I don't have any money," she said. "Besides, I've been to a lot of doctors in New York before we moved up to Maine, and no one seems to be able to get rid of them."

"Do you have them anywhere else on your body?" I asked.

She immediately started to drop her trousers. "Not necessary," I said. "Just tell me."

She sighed and pulled her sweatshirt back on. "They're everywhere—my whole body's covered with them."

The prolific nature of the sores suggested to me that the girl had a systemic staphylococcus infection, which could be life-threatening if left untreated. Whenever someone tried to get me to treat human ailments, I said that if it were my body, I'd see a physician. I repeated that advice again this time, stressing my concern that the sores could be the symptom of a serious infection. "But in the meantime, I suggest you swab some gentian violet solution on them. It's an antibacterial as well as an astringent, so it will help dry out the sores. I have a bottle in the car. It won't cure them, but it will ease itching and burning, and it shouldn't do any harm, except for maybe staining your clothing."

She thanked me profusely as we walked back out into the yard, and I started going through my usual mental list of what I had to do next. Then the young woman broke into my thoughts. "Doctor

Brown, I almost forgot—I wanted to talk to you about my turkey. I'm the group leader for the birds. We have twenty chickens, eight ducks, seven geese, and one turkey."

"Quite a menagerie," I said.

"The turkey's been awfully lame lately, and I think he's growing a tumor on his foot. Come on—I'll show you."

I followed her out to an old outbuilding. She pushed aside an old door propped over the opening. "We call him Tom," she said, smiling like a child.

Inside the shed, the young woman managed to get hold of Tom so that I could examine his foot. He squawked and wiggled, but she held her grip. After a careful inspection I assured her that Tom didn't have a tumor, but a common bird ailment known as bumble foot.

"Bumble foot?" the girl asked. "Really?"

I laughed. "Yep," I said. "Actually, it's just an abscess, an infection under the skin that causes pus to build up. As the pus dries out it forms a thick, cheesy material that presses on the nerves around it—it can be very painful. That's why Tom's favoring that foot."

I told her to hold her grip on him while I swabbed disinfectant on the turkey's foot. I injected a local anesthetic and lanced the abscess, removing about two ounces of the cheesy material. Then I packed the cavity with an antibiotic salve and bandaged the foot.

"Don't remove the bandage," I told her. "Just let it wear off as the turkey begins to walk on it again." I left her some ointment to use after the bandage had worn off.

I ran into the human Tom again as I was stowing my gear in my car. "Hey, Doctor Brown, did you happen to find anyone interested in adopting any of our puppies?"

In fact, I had completely forgotten. "No, but it always helps to have them on hand to show off," I said. "How many do you have?"

"We have three left. Our neighbor took one already," he said.

"Okay, well, if it's all right with the committee, I'd like to take them back with me, give them their shots, and let them live in the animal hospital until I can find homes for them. I have a feeling they'll be gone within a few weeks."

Tom said he'd be back in a few minutes and ran off to consult the other members of the animal-care committee. When he returned, he was carrying a box with two puppies inside. "Decided to keep one for myself," he said. "I think he kinda likes me."

I asked him to fetch his new companion so I could vaccinate him for distemper and other lethal canine diseases. When I told him I wouldn't charge him, he promptly complied. As I made my way out the dooryard with the little dogs in the back seat, I couldn't help but notice that the human plow gang had taken my advice and were plowing across the slope. I knew I had gained their confidence when two or three of them managed a faint wave as I drove out.

Over the next few months I occasionally drove by the communal farm on the way to other calls and wondered how those kids would survive a Maine winter. One day in December I found out. Tom brought his dog into my animal hospital for a rabies shot. He told me they had traded the deed to the farm for a used tractor-trailer truck. All of those who wanted to go, along with the animals, were "bookin' it" for Florida. No one stayed.

As far as I know they were the last inhabitants. The buildings gradually toppled into the ground, and alders swallowed up the open fields and lawns. People driving by a few years later would never recognize the property as a farm that, in its heyday in the 1920s, had once been thriving and beautiful.

As for the hundreds of young people who went back to nature in Maine in the 1960s and '70s, I would guess that no more than 10 percent of them stayed on, and some are still here. Given their odds, that's a lot.

18 The Racing Cure

From time to time I was called upon to treat racehorses, mostly for leg injuries and chronic lameness. Though I also treated many show horses—those that perform dressage, barrel racing, rodeo, and various gait events—as well as work horses and riding horses, working on racehorses is a specialty unto itself. In parts of the country where horse racing is big business, vets limit their practice not only to racehorses, but to particular types—thoroughbreds or standardbreds. (Both are purebred, which means their bloodlines are registered, but thoroughbreds race with saddles, and standardbreds race in harnesses.)

Although not exactly the center of the racing world, Belfast did have two of the best standardbred stables in the sport, and I would be called to care for those horses when they arrived home from racing in other parts of the country. These stables were bankrolled by two of the city's wealthiest citizens. As the saying goes, "Racing is the sport of kings."

People of lesser means get into the act, too, most hoping to strike it rich on that one-in-a-thousand chance. Rich or not, most racehorse owners are smitten by the potential of certain bloodlines. It's been my observation that people initially buy racehorses because they love the sport, and almost without exception they become addicted to it. Insidiously and progressively, winning becomes the owner's sole focus. On paper their horse looks good, and with such-and-such genetic legacy, it should be a real winner. If genetics were the only factor, it would win every time; but things rarely work out that way. If a horse does prove to be a winner, its owner becomes infected with the obsession to win more. If it's losing more often than not, the owner schemes about ways to make it the winner he knows it really is.

One owner once told me the obsession borders on insanity. "I own twenty horses so that at least one or two of them will pay for the keep of the rest of them," he admitted.

This obsession takes quite a toll on the animals. Many standardbreds in Maine come from Florida and other southern states. They slowly move up the coast, usually to Maryland, where there's another racing circuit, finally ending up in New Hampshire and Maine. By this time, many of them are physically broken down, with myriad joint and tendon problems. Some just didn't have the right stuff genetically to perform at the level they've been competing in. Others started out well and just got hurt, or old, or both.

Once injuries start to appear, less-educated racehorse owners will do almost anything, however irrational, to find a cure. Many are highly superstitious. If they thought distilled water poured over the horse's head would help it run faster, they would corner the market on distilled water. The way news travels in the racehorse fraternity, their colleagues would be doing the same thing the minute the word got out. If so-and-so put a red-colored liniment on his horse's sore tendons and that animal won a race, then the next day every medicine cabinet in the stable would contain a bottle of red-colored liniment. If the same ingredients were formulated in a blue-colored base, that liniment would be ignored.

Racehorses are, of course, athletes, and the worst thing an owner wants to hear is that it's time to put a horse out to pasture. When dealing with a chronic injury to their legs, I'd use the analogy of great human athletes. "You know, it's like putting him on the disabled list," I'd say in my most persuasive tones.

But the spiel had little effect on the most addicted owners. Granted, they were in the minority, but I tend to remember them the most. They tend to be the least well off and most financially desperate, a combination that doesn't bode well for the health of a horse. These are the folks who have an almost feverish faith that their horse can be patched together somehow, even if it means resorting to some barbaric quackery.

Even the more reasonable owners who conform to modern-day treatments and accept what a professional vet tells them still don't like to hear that a chronic injury is putting an end to their animal's career. When it comes to the end of their racing years, the majority of racehorses end up being slaughtered for dog food and for human consumption in foreign markets.

Racehorses take a terrible pounding, and it's a wonder that their legs and joints don't wear out much sooner than they do. Their owners blame everything—weather, tides, and mothers-in-law—if their horses lose a race, and the less-educated owners don't even consider physical and mental ability. I have always contended that genetics determines the speed capability of all racing animals, in all races. Given the same training, conditioning, and nutrition, I believe genetics makes the difference. To make an automotive analogy: rarely, if ever, can a Chevy beat a Lamborghini in a mile race.

This was the backdrop to the visit I received one late summer afternoon during my regular office hours. Out of the window I noticed a pickup pull into the parking lot towing a double horse trailer. Two men exited the pickup, and after a few minutes, my assistant let me know they were waiting their turn to see me. When he announced them I nodded in recognition. They were two brothers who had had several dogs over the years and owed me a fair amount of money, but I'd never known them to own horses. I no longer recall their names.

Half an hour later I ushered them into the exam room. As usual they were dressed in well-worn jeans and T-shirts that had seen a day or two of heavy use in some physical occupation, maybe auto mechanic or woodlot work. I didn't really know what they did, but it was clearly not too lucrative.

"How can I help you guys today?" I asked.

They told me they'd purchased two standardbred racehorses and asked me to come out and look at them, as the horses were having the usual difficulties with legs and joints. I told them to bring the horses around to the corral beside the parking lot. When I met them out there I encountered a pair of sad, broken-down horses who should have been retired several years earlier. After circling every track from Florida to New Hampshire, these fellows had convinced an unsuspecting uncle to bankroll them in their get-rich-quick venture. The uncle, though hardly wealthy, was apparently better off than his two nephews. They told me they had heard from their buddies in the racing circuit that there was a new injection to rejuvenate weary racehorses, and this is what they were hoping I would provide.

They had the poor old nags' health records with them. Reviewing them I saw that from an early age to the present they had been given everything from thyroid to thiamine. They had received countless injections of steroids in their knees, hocks, and ankles. Worse, they had been tortured with outright quackery, such as firing (burning the tendons with a hot iron) and blistering (applying an acid that causes the skin to blister, also to "treat" tendonitis). I kept my opinions to myself. It was clear that they were financially desperate and hell-bent on pursuing this fountain-of-youth cure they'd heard about. In such cases I've learned to save my breath and give the animal a placebo, knowing that the clients will just keep driving until someone did something more drastic.

So I convinced them to let me give each horse an injection of B12, iron, and other B-complex vitamins, as I knew it would do no harm. But I did so with the stipulation that if the nags finished last or next-to-last in their scheduled race at the upcoming state fair, they would consider selling them to me for their delinquent vet bills plus the day's services. Also I extracted a promise from each that they wouldn't bet a nickel on either horse. After several minutes of conversation back and forth, they reluctantly agreed to my terms. When they left, they were considerably deflated, having read between the lines that they'd been swindled by whoever had unloaded the horses on them.

The brothers raced their horses the following Saturday at the Bangor State Fair. They finished the race last and next-to-last, respectively. I now owned two broken-down racehorses, surrendered without a struggle. In negotiating this deal I had a plan in mind for these two horses. It involved two teenage girls from two different families, each of whom lived on a dairy farm and had wanted a horse more than anything. They didn't know each other, but their yearning for a horse was so similar that in my mind, they shared a sort of kinship.

I boarded the horses at a stable in the country and continued to nurture their general health. They responded to tender loving care and were free of any further harmful "treatments." Within a few months they converted to saddles readily, and the stable owner started training them to become saddle horses. After about six months they were rehabilitated and ready to meet their two future owners.

I knew that each of the girls came to this stable occasionally to ride for a few hours. One day I called them both and told them, "Hey, I hear you want a horse, is that right?" It was great fun to listen to their shrieks when they realized I wasn't just teasing them. "Go on out to the stable and meet your new horse," I told each one in turn.

I wasn't there, but the stable owner reported that it was love at first sight. The first girl arrived and chose the horse that approached her first, so no conflict arose when the second girl showed up. They bonded with the horses fast, and after several days of riding at the stable, the two girls had also bonded with each other. When the stable owner felt that all was well, she trailered the horses and dropped them off at their new homes.

Over the years I watched the bond grow even stronger between these two horses and their owners. As I recall, one horse lived eight and the other twelve years before dying peaceful, natural deaths after years of relatively good health. They had found that fountain of youth after all.

19

The Three Sisters

Over the years I treated sheep from time to time. Though I knew several farmers who kept small herds of goats, very few Maine farmers raised sheep for a living because of the large acreage they require for grazing. As a result, sheep were usually a sideline for farmers who liked to raise them for wool, meat, or both.

One of these was Charlie McKinney, a chicken farmer on the outskirts of Belfast who kept about fifty purebred North Country Cheviots, strictly out of love for the species. He once told me that there was nothing as comforting in life as holding a newborn lamb. Charlie spent many hours watching his lambs cavorting when they had been first turned out to pasture. With his love of the young animals, it was heartbreaking for him to sell them for meat in the spring lamb market. He literally wept when they were loaded into the truck, bound for slaughter. I once asked him why he put himself through this torture, and he replied that he needed to justify the expense of keeping the animals to his wife.

Charlie called me at five o'clock one morning in May. "Doc," he said, "my favorite ewe has been in labor since about two this morning. It's her first birth and I can tell she needs your help."

I arrived in about twenty minutes and was presented with the tail of the lamb protruding from the birth canal—the sign of a breach. I corrected it without much trouble and soon delivered a healthy female lamb. I went back in to fish for another, as twinning is common in ewes. Sure enough, minutes later I extracted a healthy twin sister. Then I probed the mother's uterus for the afterbirth—the placenta that nourishes the fetus in the womb—to clean it out and insert antibiotic boluses to prevent infection, a routine procedure.

To my amazement I felt the head of a third lamb. Its head and neck were folded back on its flank, a dystokia, or abnormal birth position. The mother never could have delivered this lamb—it was so out of position that a normal birth was physically impossible. Without saying anything to Charlie, I put my middle finger in the mouth of the fetus and gently went to one corner of the mouth and gently waved the head dorsally, at the same time bringing the head and neck up and then forward into the birth canal. The little lamb was sucking my finger the whole time. Once the head and neck were pointed toward the outside, I let nature take over. Squeezing through the birth canal would compress the newborn's chest cavity, expelling the viscous, honeylike, mucoid fluid in the lungs that is present at birth in all vertebrates, including humans. This heavy fluid keeps the lungs properly shaped during the months of development inside the womb, but life outside requires the infant's air passages to be clear.

Now that the final newborn was free of the mother, the umbilical artery would be able to close off as the artery muscle, called the intima, constricted. Because the umbilical artery is so large and carries so much blood, the mother would soon bleed to death if that vessel didn't squeeze shut to prevent blood from streaming out. In the horse, for example, the umbilical arteries are at least an inch in diameter—large enough to cause a full-grown mare to bleed to death in minutes. But the artery shuts right off after delivery, so that not another drop of blood escapes. Tell me that's not a miracle.

Once the fetus has entered the world, the arteries that brought the lifeblood to the little creature become the so-called round ligaments of the bladder, and the umbilical vein becomes the falciform ligament of the liver. More miracles.

Charlie watched, his mouth agape, as I helped the third lamb, also female, into the world and placed her in front of her mother, who was busy cleaning off her sisters.

"I'll be darned, Doc," Charlie exclaimed. "I've never had triplets before."

I smiled. He sounded like a proud father. "Well, it's very unusual," I agreed. "And they're all healthy, too."

Seeing her third baby, the mother turned her attention to the newest lamb, licking her from stem to stern. The first two were already struggling to try out their legs. Like all four-legged animals, newborn lambs begin trying to get up immediately, usually succeeding within five minutes. The first two triplets found their feet, and, though wobbly, began nosing for milk at their mother's teats. The third sister, now suitably clean, also stumbled to her feet.

At this point Charlie, who was so excited that he could hardly speak, ran to the house to get his Polaroid camera. I watched the newest lamb push between the others and struggle for milk. The others were slightly larger and shouldered her out of the way. Charlie bounded back, out of breath, and snapped picture after picture. He took one of me holding the triplets, then I took one of him holding them. They were indeed beautiful, and a real handful, too, squirming against our arms and bellies while bleating for their mother.

"Better keep an eye on that third one," I cautioned. "She's getting the elbow."

"Yeah, I noticed that, too," he said. "Don't worry, Doc."

I got ready to depart for my next call, washing up with tamed iodine, a potent but less irritating disinfectant than the classic iodine solutions. I then scrubbed down my boots, picked up my gear, and headed for the car. I said goodbye to Charlie, who was still snapping away, but he barely noticed my leaving. I had just closed my car door and started the engine when Charlie came rushing up and shouted, "Wait!"

"What's up, Charlie?" I asked through the open window.

"I got so taken with the triplets that I almost forgot—I need some of that vaccine that prevents my best lambs from dying when they are turned out to pasture the first time. I can't pronounce it, but you know what I mean."

"Yep. Clostridium perfringens Type D vaccine. It protects your lambs from a bacterium that causes sudden death in the healthiest lambs."

"That's the stuff. I've never quite understood why the healthiest ones are the most susceptible, though."

After rummaging in my back seat and not finding the vaccine, I shut off the engine and walked to my trunk to see if I had some in there. "Well, it seems peculiar until you realize that the healthiest lambs are the best eaters, and if the grass they're eating has a lot of this deadly Clostridium perfringens on it, they're the first to fill themselves full of it."

"Huh," Charlie nodded. "Makes more sense now. Are lambs the only ones that get it?"

"No. The Clostridium bacteria family is a pretty bad bunch. Botulism, for example, is caused by Clostridium botulinum. Another one is tetanus, caused by Clostridium tetani. And there's one called Clostridium chauvoei that causes blackleg in cattle. The list goes on."

"So the kind the lambs get, does it live in the grass, or what?" Charlie asked.

"Bacteria that cause Clostridium diseases are found both in soil and in the intestinal tract. They're usually harmless. But under the right conditions, they grow rapidly and release toxins that destroy tissue. If those little lambs get filled up with milk or grass and their digestion slows down, that could allow the bacteria to multiply to a toxic level. That's why the disease is sometimes called overeating disease."

Charlie looked grave, probably picturing one of his new babies laid out on the grass.

"It's another name for pulpy kidney disease, because Clostridium Type D toxins destroy the kidneys," I rattled on. "If you open up a dead animal for a postmortem the kidneys look very pale, as though they've been blanched."

Charlie put his hand up. "OK, Doc. Enough."

"Sorry, Charlie. I tend to get carried away." I told him I didn't have any of the vaccine with me and to drop by the office when he was in town.

Charlie, no doubt spurred on by my horrifying Clostridium lecture, came by the next day to pick up the vaccine. "I'm raising that last triplet on a bottle," he told me, when I asked how the new lambs were getting on. "The other two have taken all their mother's milk, greedy little devils. Except I let the third one suckle once a day to get those antibodies that you said they need in the first milk from their mothers—what is it, colostrum?"

"Yep, that's it. So, Charlie, what did you name these three?" He named every sheep he owned.

"Well, I was going to call them after my three aunts, but I don't like one of them very much," he said. "Then this other idea just flashed into my mind, so I went with it. They're called Hope, Faith, and Charity."

I had to agree that these perfectly captured the three lovely sisters.

20

Sally's Knee

I was doing surgery one afternoon when my assistant in the front office appeared in the doorway and asked if I could see Bessie Goodhue. "Sure, Bessie's got a strong stomach," I said. "Send her on out."

Bessie appeared a moment later, standing at the foot of the operating table while I extracted the ruptured spleen of a dog that had been hit by a car. She watched with curiosity, showing no revulsion. Her face was like leather, lined with deep wrinkles. That skin, and the fact that she had lost all her teeth years ago, made her look much older than her age, which I would have been hard-pressed to guess. She had a masculine demeanor, muscular arms, and stood no taller than five-feet-two. She wore her gray hair cut short, often under a baseball cap, and more than once had been mistaken for a man by people who worked for me.

Bessie had lived her life so far with her brother, Ralph, on a little farm that had been left to them on their mother's passing. Their father, also deceased, had kept Guernsey cows and peddled milk in a horse and buggy for years after most people were using cars. He was one of the first in the area to pasteurize his milk, which he accomplished with a second-hand pasteurizer purchased from a local dairy when it went out of business. Bessie and Ralph were not educated and had worked on the farm from dawn to dark every day since they were five years old. Their little side-hill farm had many acres of pulpwood, mostly poplar, and in addition to their many other chores they peeled pulpwood and sold it to a pulp mill in Winslow. They peeled, or stripped the bark, with hand tools, a very punishing and physical task, especially during mosquito and black-fly season. One look at Bessie's hands told the story of her life: they were barked up, callused, and nicked with small cuts and bruises.

Bessie walked to Belfast from their farm in Swanville, sixteen miles round trip, to make payments on her bills in town and sometimes pick up a few groceries. Why she didn't use the mail for this task I never asked; I assumed it was because she didn't trust the government. Maine has its share of people who choose to live isolated rural lives and harbor suspicion of the government, so this wasn't terribly exceptional, although her walking regimen was. Though poor, she was religious about paying her bills, and normally she paid me fifty cents a week.

She incurred her vet bills when she asked me to treat her goats from time to time. Since her childhood she'd had an undying love of goats. Their father gave her a pregnant nanny on her sixth birthday, and over the years she grew her ownership to over thirty fine milking goats. They were the love of her life. Her goat milk was widely sought by people up and down the coast and into central Maine, but Bessie failed to raise her prices over the years enough to keep her out of abject poverty. Once in my presence she said, "I paid you a dollar this week because I sold a pair of kids to my neighbors yesterday. Their children want to start a milking herd so when I can't continue, they'll get my milk customers."

Today she had a medical question for me, having already paid my assistant her fifty cents for that week. She reached into her pocket and pulled out an old glass saltshaker, which contained what looked like a flat, short piece of string. "What's this in here, Doc?"

"That's a tapeworm, Bess."

"I'll be damned," she said. "I've heard of these, but I've never seen one. I found it hanging out the rectum of one of my yearlings. What do I do to get rid of 'em?"

"I'll put you up some medicine in a few minutes when I'm done with this. That should take care of it."

She jerked her head in the direction of my surgical case. "What's that thing you just took out of that dog?"

"That's his spleen, Bess. He got run over by a car about half an hour ago, and his spleen got broken. See, it's in two pieces."

"Christ, Doc, you can sure get into a lot of messes. You're not spleeny, that's for sure." She laughed at her quip. "Spleeny" is a Maine expression for squeamish.

I finished the surgery and put up her tapeworm medication. As I was about to leave on evening barn rounds and was going by her house, I asked her if she'd like a ride home. She allowed that it would be mighty nice of me.

Off we went, and for the entire eight miles she grilled me about goat problems. I pulled into her yard, and she thanked me for the ride. I turned the car around and started to leave when I heard a banging on the trunk. I rolled down the window.

"Damn, I nearly forgot. While you're here, would you look at Sally's knee? It's got a big bunch on it."

Sally was one of her favorites. Bess led Sally out to the milking table—a station that allows the goat to eat while being milked at a comfortable height for the milker—and I palpated a grapefruit-sized bunch on one of the goat's front knees. Sally, preoccupied with the extra round at the feeding station, barely flinched. The bunch had a soft center full of fluid.

"I'm going to have to open it," I told her.

"Do what you got to do, Doc."

So I went to the car and got what I needed. I injected the knee with a local anesthetic, waited about fifteen minutes for it to take effect, and lanced the bunch with a scalpel. As I'd expected, about a half pint of pus the consistency of mayonnaise spurted out, and I was ready with some gauze and paper towels. The stuff smelled about like I'd expected, too, like rotted meat. I washed out the wound with disinfectant solution, and while I was swabbing out the inside I discovered the end of a piece of wire. I gently tugged at it, kept on tugging, and by the time I removed the entire wire it was a good five inches long.

"Christ, now I've seen it all!" Bessie exclaimed. "No wonder that bunch was so big. And Sally hardly complained."

I instructed her about the aftercare, gave her some medication I called healing oil, a concoction we mixed at the office made up of urea, sulfanilamide, and nitrafurizone. It promoted rapid healing for all kinds of wounds. By the time I'd gotten the medications together she'd conned me into trimming the rear feet of her billy goat. After doing that in short order, I washed up, put my gear away, and headed off for my evening barn rounds before Bessie thought of any other goat maladies.

Several weeks later Bessie stopped in to pay her weekly installment on the vet bill. I wasn't in at the time, and when I returned I saw that she had left a photo of Sally's knees. On the back of the photo, she'd written, "Doc, I bet you can't tell which knee it was." In fact, I couldn't.

The note went on: "I took the wire to my Grange meeting tonight and told the story about where it come from. After I was done, almost everyone was feeling their knees."

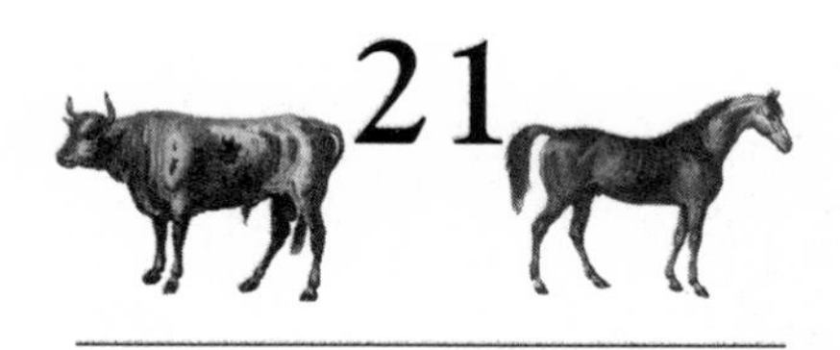

21

Loco Cow, Love-Struck Moose

"Doc," said Freddie Benton over the phone, "I got a cow that's gone loco, running in circles down in the meadow. She's spun herself just about out, and now she's lying there having fits. What in hell you think's the matter with her?"

"Pretty hard to tell over the phone, Fred, but I can probably get out there within an hour or two if you want."

"Thanks, Doc. If you can get here sooner, that'd be even better. She's in bad shape and it's gonna be dark before long."

"I'll come as soon as I can, Freddie."

"Oh, Doc?"

"Yes?"

"You got a shotgun?"

"Why?"

"There's a bull moose up in that meadow and he put me right up a tree yesterday. I got a deer rifle, but might be good to have some kinda cannon to fend him off if need be."

"All right, I'll throw my Browning automatic into the car."

It was a lovely September day, the maples were just beginning to turn gold, and Maine's moose were coming into rut, or mating season. Freddie's moose encounter was the kind of nightmare you heard about every so often. At this time of year, the male or bull moose shed the velvet from their impressive antlers—their "rack" can reach a width of six feet and weigh more than fifty pounds—and gain bulk in their necks and shoulders, hormonally primed to fend off competition for females. In this territorial state, they were known to charge anything that crossed their paths. Luckily, their paths were usually well out of the way of people. I had no doubt that Freddie was telling the truth, but I wasn't afraid that this rare occurrence would repeat itself.

When I arrived at Freddie's place in Waldo I found him much more nervous and talkative than usual. He stood beside me as I gathered my equipment from the trunk of my car, retelling the story about the moose, the saliva building up in the corner of his mouth as he re-enacted the whole scene. "I was just walkin' up through there, mindin' my own business on that old woods road. That bull moose, he must a weighed at least a ton, took up the whole road. He stepped right out thirty feet from me. He kind a looked me over for a couple a seconds and lowered his head, then he started after me." Freddie paused to catch his breath.

"Let's start walking, Fred," I said. "Like you said, we're losing the light. Why don't you carry the gun? All I've got for shells are number six birdshot. Load her up. She holds five. If he bothers us you can let those rip over his head—that ought to scare him pretty well."

"That's about all that birdshot'd do, except maybe sting him a little," Fred allowed.

After loading the shells, Fred took up the story again as we set out down the woods road toward the meadow. "I started for the nearest tree, which happened to be a fir about eight inches thick. I broke off the first four or five branches climbin' up that son of a gun just to get high enough to look down at those antlers. Then he started to paw and roar, and he charged that tree so hard he shook my false teeth outta my mouth. I'll tell ya, I wasn't about to go down and pick 'em up. That crazy moose kept me up that tree for nearly an hour."

"Why did he finally leave?" I asked.

"My neighbor's dog came crashin' through there flat out, just a ki-yi-in' after a rabbit. That distracted him, and he finally took off down the road toward the meadow. So you see what we're up against here, Doc."

Fred kept a sharp eye out for the moose as we walked along. Every so often he'd crouch down and peer through the brush, and a snap of a twig brought the shotgun to his shoulder, ready for his formidable opponent.

Despite these distractions, I did manage to get a little history on the cow. She'd been feeding in that meadow for about two days, but Fred insisted that she'd eaten there for the last four years and never had a sick day in her life. I'd long known that Freddie wasn't the best feeder in the world, and his pasture had been grazed to the ground. That explained the bony condition of the cow that lay jerking in the grass ahead of us.

It was obvious that the cow had a neurological problem: her head was pulled back, her eyes rolled up into her head, and all four legs were paddling as if to escape a swarm of hornets. The fight-or-flight part of her brain was definitely in charge. The poor thing was exhausting herself, and I was worried that her heart wouldn't stand the exertion much longer. Before searching any further for a diagnosis I gave her a dose of mild anesthetic, which stopped the convulsions. I took her temperature, checked her over with my stethoscope, and let my mind go over the possibilities. My examination ruled out tetanus, rabies, listeriosis, and a host of infectious diseases, so I started playing a hunch.

"Fred, does she have the run of this whole meadow up to those old stone walls up there?"

"Sure, why?"

"Well, I've got a feeling that she might've gotten into something—a plant such as a fern or something else toxic to cows."

"Huh," Freddie said, the shotgun crooked over his arm. "Never heard of such a thing."

"Let me just take a blood sample here." Withdrawing some arterial blood into a test tube, I was startled to see that it lacked its normal color: instead of a bright, robust red, it was a dark brown—more like the blood in veins, which carry oxygen-depleted blood back to the heart and lungs to replenish the oxygen supply. Maybe something was tying up her hemoglobin, the protein in red blood cells that picks up oxygen in the lungs and delivers it to the surrounding tissues.

I began to think my intuition about a poisonous plant might be right.

I left Fred with the patient and struck out over the meadow to see what plants were growing there that may have put her into such a state. An undergraduate botany course, followed by a course in poisonous plants in vet school, had taught me that about eighty plants are potentially lethal to both people and animals. Some of those, such as digitalis, commonly known as foxglove, also have medicinal uses. Whether a plant substance is a bane or a balm depends largely on the concentration and the amount ingested.

As I followed a serpentine course through the meadow I saw many possible culprits: bracken fern, skunk cabbage, buttercups, purple nightshade. But they showed no sign of having been grazed. Unless the pastures are completely stripped of grass, animals usually don't resort to eating poisonous plants. But then, grazing typically thins out in fall, when some poisonous plant species contain their most lethal levels of toxins.

I reached the stone wall on the crest of the meadow. Up to the wall, meadow plants prevailed. On the other side of it, a young forest was overtaking a former pasture, shading both sides of the wall and creating a habitat for ferns, moss, and lichen. I walked slowly along the old stone edifice, trying to observe everything that grew. Then I spotted it: a patch of chokecherries waved in the breeze on the woods side of the wall, well within reach of a cow. In fact, as I looked closely, I discerned a browse line where she had nibbled fruit from several branches.

Chokecherries (*Prunus virginiana*) contain prussic acid, a form of cyanide. When consumed in quantity, the acid prevents the blood's hemoglobin from carrying oxygen. This fit my hypothesis and would account for the dark color of her arterial blood.

I rushed back to my patient and explained the diagnosis to Fred as I injected an antidote, a mixture of sodium nitrate and sodium thiosulfate, which reverses the absorption of prussic acid. I always carried antidotes to common poisons in my grip, especially in the fall.

"Well, I'll be," Fred mused.

As I was giving the last injection, he asked, "Hey, Doc, while you're here you wouldn't pinch off a bull for me, would ya? He's gettin' a mite frisky. Should a called you awhile back about him, just never got to it."

"How old is he, Fred?"

"He must be pretty near twenty-one months."

My heart fell, knowing full well that he was not a candidate for "pinching"—performing an external vasectomy by crushing the vas deferens with an emasculatome. That should be done between four and eight months of age, when the testicles are still growing. After that point, surgical castration using a local anesthetic is the only option, because pinching would cause the animal too much pain, and a bull that size in a lot of pain can be hard to control.

It would take several hours before we could tell if the cow was responding to the medication, so we left our patient and started back along the woods road to the barn. I was explaining why I couldn't pinch the bull when we heard a sudden loud crack, like the sound of a large branch breaking, in the woods on the right side of the narrow road.

"Get down, Doc!" Fred hissed.

Crouching in the brush on the left side of the road, we spotted a bull moose cutting through the trees ahead. He was moving at a trot directly in front of us, about fifty yards away. With front legs more than six feet long, he effortlessly bolted a ditch and hit the roadbed without breaking stride. Then he turned and headed straight for us. Moose have weak eyes, but their hearing is keen, as is their sense of smell. Our cover was blown.

Fred was still carrying my shotgun. As the moose picked up his pace to a canter, Fred whipped the gun to his shoulder and squinted down the barrel. About fifty feet from our position, the big male slowed to a stop and lowered his head, displaying a five-foot spread of antlers. Pawing the ground furiously, he let loose a bellow that stood my neck hairs on end. Then he charged.

Fred didn't wait another second. He dropped to one knee, raised the shotgun, and fired two quick shots over the bull's head. It did the trick: the moose made a ninety-degree turn and thundered into the woods. We could hear him crashing through the underbrush as we continued up to the barn, a bit unnerved.

"Hoo, baby!" Freddie said.

"Nice job," I said, letting out a big breath. "I can see why you would've been a little nervous up in that fir tree."

"You know, Doc, that 'tweren't the same bull moose. I'm certain of it. This one here's a dite smaller and a lot darker. Just as mad though."

"You really think so?" I said.

"I'd swear on it," he allowed.

When we arrived at the barn, Freddie led me to my next patient, a big, rangy Holstein bull, tethered by a neck chain to a homemade wooden stanchion next to the rear wall.

"There's Brutus," Freddie announced. Though only twenty months old, Brutus could easily have passed for a four-year-old. He was fully matured, both sexually and in stature. "You gotta get behind him to do it, don'tcha?" Fred asked.

"Of course," I said. "Why?"

"Well, I tell ya, he's mighty light-footed back there. He'd kick the stars out of heaven if he got a chance. But I s'pose you have a way of handlin' these fellas."

Yeah, I thought, when they're four months old. "Hey, Fred," I said aloud, "I see you've got a pair of kicking chains over there. I'm gonna try to slide those over his hocks before he knocks my teeth out. Then you're gonna help me keep his back feet on the floor long enough for me to push up hard on his tail and block the nerves near his scrotum with a shot of local anesthetic—same stuff a dentist uses to numb your teeth."

Freddie nodded without enthusiasm.

"After we get his scrotum numb, I'm going to cut the testicles out of him. But before we do a thing, I'm going back out to my car and get a nose lead. I'll clamp that in his nostrils and have you hold it off to the left. That'll keep his big rear end from crushing me against that wall and distract him while I'm trying to get these kicking chains on him."

Just then the bull snorted and tossed his huge head. Fred jumped involuntarily.

I retrieved my nose lead—a metal tong that I could clip into the bull's nostrils, fastened to a strong rope—and returned to the barn. Freddie had fetched the kicking chains. Carrying the nose lead in my left hand, I approached the bull from the front and slung my right arm around his head. With three shakes, he slammed me against the wall and drove my head into a beam so hard I blacked out. I recovered to find myself sprawled on the floor at Fred's feet.

"Damn, Doc," I heard him say through the fog. "Your head should be split open like a cantaloupe the way he flung you into that beam. Sounded just like a razor strap comin' down on a wet board."

Fred helped me to my feet. After my vision cleared, I went at Brutus again. This time I was mad. After three or four minutes I succeeded in putting in the nose tong. I threw the end of the rope up over the old dusty beam and handed the other end to Fred, instructing him to keep it taut. Then, grabbing the kicking chains—actually a single chain with clasps—I went for Brutus's back end. I'd learned over the years the closer you stay to an animal, the less chance you have of getting hurt. So with the kicking chains in my right hand I drove my left shoulder into his flank with all the strength I possessed, hanging in there as Brutus bobbed up and down and kicked violently. In an odd way it was like dancing, except I didn't know the steps, and Brutus was definitely leading.

After a few minutes I succeeded in getting the chain around one leg, clasping it above the hock. I was about to clasp the second leg when Brutus kicked back with both feet so violently that the chains went flying into Freddie's manure-specked window, shattering it to smithereens.

"Don't worry, Doc," said Freddie, spitting a wad of chewing tobacco onto the floor, "I don't need that window this time of the year. So as long as nobody got hurt, pay it no mind."

I checked myself and glanced at Brutus. Both of us had escaped the shower of glass. Storming outside, I picked up the chains where they'd fallen and returned to my adversary. I was mad before, but my anger had gone up a notch.

"You're not getting rid of me, Brutus," I said.

"Atta boy, Doc," Freddie said.

With raw determination I succeeded in clipping the chains onto his hocks, which hobbled him enough to narrow his striking range. He snorted and flailed his hooves at me as I went behind him and started to inject his spermatic cords, where the spermatic nerve and six other structures are located. We rocked and rolled awhile, but I succeeded in getting an injection into him.

While we waited for the anesthesia to take effect, a cattle dealer and friend of Freddie's wandered in and strolled toward us. After greeting us he studied the situation for a minute.

"Freddie, you ain't gonna castrate that bull, are ya?" he said.

"No, Thurmond, we're making banana splits," Freddie replied.

The dealer laughed. "Okay, okay. What I mean is, what the hell are you doin' that for?" he asked.

"I want to put this fella in the freezer as soon as he gains some more weight and gets the bull taste out of his meat. Should make good eatin'," Fred replied.

Thurmond tilted his head to one side. "He's out of Willow, ain't he?" he asked.

"Yep. His daddy is Willow and his ma is Creamaster Deluxe. His daughter's my best milker," Freddie replied.

"Well, by God, Fred, I've got a buyer for a critter with that pedigree. Wants to cross Willow blood with some Stallmaster daughters he's got," Thurmond said.

Freddie perked up, but pretended he wasn't sure what he was going to do. They haggled for two or three minutes, and after I'd nearly lost life and limb wrangling Brutus, the dealer arranged to pick up the damned bull the next day. "I'll bring you a check then," he promised.

If I was the sputtering sort, I would have sputtered. Instead, I gathered my gear and cursed silently as Freddie walked Thurmond out to the dooryard.

Fred met me at my car as I stowed the nose lead. "Come on," I said, "let's go check on that cow. I have some other calls to make."

Freddie grabbed my shotgun and we headed down the woods road to the meadow. Freddie was quiet, sensing I was none too pleased. He returned to lookout duty, glancing here and there for signs of the moose, but we arrived in the meadow uneventfully. There we were amazed to find that our patient had regained her feet only an hour after receiving treatment. Although weak and wobbling, she was making a fast recovery. My mood lifted.

"What do you say, Fred—want to get this beauty back to the barn?"

"I don't have a lead on me," Fred replied.

I wore coveralls and had no belt, but I saw that Fred was wearing one. "Why don't you take off your belt and put it around her neck?" I suggested. "I'll grip her tail and help move her along."

"Okay," he agreed, "but you'll have to carry the shotgun. Without my belt, I'm gonna need my other hand to hold my pants up."

We proceeded with the plan, coaxing the cow through the

meadow and onto the woods road, stopping from time to time to give her a rest and allow Fred to hitch up his trousers. We had stopped about three-quarters of the way to the barn when we heard a damnable ruckus just off to the left of the road about forty yards away. Gazing into a clearing, we made out the forms of two enormous bull moose locked in battle. It was a rare and unforgettable sight, but we didn't stay long, fearing to divert their attention.

"There," Freddie exclaimed, "I told you there was two of 'em! Guess I got caught right in the middle of a moose fight."

"I reckon so, Fred." We continued our trek to the barn, secured the cow in the tie-up, and gave her a pail of fresh tepid water. She gulped it down, and we offered her a little grain, which she ate with relish.

"Well, now, it looks like she's on the upswing," I said. I instructed Fred about her aftercare and advised him to cut the chokecherries back out of reach of the cattle. Then he asked what he owed.

"Tell you in a minute," I answered, reaching for my billing pad. As I totted up the charges, Fred cleared his throat.

"Uh, you aren't gonna charge for castrating that bull, are ya? I mean, 'cause I didn't have it done."

"Heavens, no, Fred. I wouldn't do that." I handed the bill to Fred and he studied it very carefully.

"Boy, that's more than reasonable, Doc, but I ain't got a red cent 'til I sell Brutus, and to tell you the truth, that money's already spoke for. Would you settle for some damned good potatoes?"

"All right, Freddie."

"Good enough," he said, relief spreading over his face. "I'll bring down a bushel on Monday morning."

As I started up my car, Fred's brow wrinkled in thought for a moment and then he said, "Hey, Doc, the thought just struck me—that cow was in heat yesterday when she took sick. Do you suppose those two moose was fightin' over her?"

I considered it, vaguely remembering tales of cows pining for moose and vice versa. "Well, I guess it's possible, Fred. But alas, we'll never know."

"Huh," he said. "Okay, then, see ya Monday with them potatoes. If it don't rain."

Epilogue

When I began to face the thought that I'd have to retire much earlier than anticipated, I mulled over how to spend the next phase of my life. Having grown up on a farm and spent so much time among farmers in my practice, I decided to give it a try myself. In 1976 I bought my grandfather's farm, which had passed to my aunt and uncle. They had sold off their Holsteins long ago, but the milking parlor and other equipment were still there. With my aunt deceased and my uncle in a nursing home, I thought, well, if it doesn't work out, at least I'll be keeping the property in the family.

So I bought a herd of cows in 1977 and stocked the farm. But I jumped the gun, because I was still practicing in Belfast, so I had to hire people to run the place. Of course, I hardly ever got over there.

My mother, who worked in the vet hospital keeping books and making appointments, warned me that my farming venture was going to be a disaster. Seeing the expenses mount as she wrote the checks, she wasn't shy about reminding me of my folly. After two more years I could no longer ignore my health problems, and I sold the practice in 1978. Only then did I return to the farm my maternal grandfather had worked during my childhood, a mile from my parents' farm. By this time I was divorced, and I moved in on my own.

Almost immediately, I saw the overwhelming nature of what I had taken on. Even the constant exposure to the exhausting regimen of all the farmers I worked for had not prevented denial from setting in. When the cows needed tending and milking in the dark winter mornings, when a window needed replacing or a tractor fixing or a milking machine cleaned, there was no recourse but to deal with it—immediately. My fantasy quickly melted away. I knew that I couldn't physically endure the rigorous demands of the farming life.

So I sold the cattle and simply lived there, taking care of the house and four outbuildings, which proved to be job enough. I taught myself a little plumbing, a little electrical work, and a little automechanicing. Between maintaining the buildings and keeping up with numerous small jobs that needed doing, I was almost fully employed.

In August 2003 I suffered a stroke, and now I can do far less than I want to do. It's frustrating sometimes, but it has made me appreciate my general good health during my twenty-three working years as a veterinarian and the twenty-five years before that. I was able to put three children through college, and now I'm contributing financial support to my four grandchildren for their educations. I have no complaints. My life has been rich.

So You Want to Be a Vet?

Most successful veterinary students have been better-than-average students from kindergarten through pre-medical college course work. Though most veterinary students enter their training with a bachelor of science degree, some have earned master of science degrees, and a very few have Ph.D.s in science. Regardless of their degrees, by this time all have learned how to study efficiently, stay focused, and keep their eyes on the prize. For the academically gifted, some colleges and universities offer two-year pre-veterinary programs, which shorten veterinary training from seven or eight years to six. The American Veterinary Medicine Association website, www.avma.org, contains information about the twenty-eight veterinary programs in the United States.

When I attended vet school in the early 1950s, there were only sixteen in the country, and the GI bill was flooding the universities with World War II veterans. It was nine times more difficult to be accepted to a veterinary medical school as it was to get into a school of human medicine. While this was true in every endeavor, it was particularly true with professional schools. Naturally, veterans—our national heroes—were given first dibs.

When I finally got an interview at Cornell University College of Veterinary Medicine (New England did not then have a veterinary school) I was twenty years old, and the admissions committee told me there was only one opening left. They still had two hundred and fifty more applications to review, out of thirteen hundred in all, from which to select a class of fifty students for the DVM program. Of the fifty students taken per class, only about eight or nine were chosen from outside of New York State, and most held a bachelor's degree. No doubt my brother Phil's extraordinary track record before me helped enormously, as did being raised on a farm. Also, I told the admissions board that four years hence I would be practicing large-animal medicine, and at that time the number of large-animal practitioners was dwindling as more and more vets chose small-animal practice.

For me, the anxiety didn't end upon being accepted. Despite public perception to the contrary, a DVM (Doctor of Veterinary Medicine) degree is as difficult to obtain as an MD (Doctor of Medicine) degree, if not more so. Even today, few people outside the profession know that vets put in six to eight years of extremely challenging academic training. The rigor of veterinary training earns a great deal of respect from students in other professional colleges, however. When other Cornell students asked what school I attended and I told them, they invariably commented that I'd chosen the toughest college at the university. Such comments don't soothe your nerves if you're in your first year.

Some students will fall along the wayside for various reasons—a couple will fail academically, a couple more will experience too much stress to continue, a few more will have girlfriends or boyfriends who pull them off course. Anyone stepping up to receive a degree on graduation day has passed through one of the toughest programs that exist.

One other thing: For those who set out to practice large-animal medicine, it's crucial to be in top physical condition, as at times the animals you deal with will test your brawn as well as your brains. I had been very athletic in my younger days. In my four years of high school I played three sports, earning ten varsity letters. In college I played basketball and one year of football, and after vet school I continued to keep myself physically fit.

Domesticated animals will always need medical attention, and veterinary medicine will always need well-trained, compassionate people. Anyone who pursues it has my admiration and best wishes for a rewarding and satisfying career. If you succeed in receiving your DVM and enter practice, keep in mind an old saying that has served me well: Never be the first to have tried, but never be the last to cast the old aside. And try, most of all, to do no harm.